Occurrence Survey of Boron and Hexavalent Chromium

Awwa Research Foundation

About the Awwa Research Foundation

The Awwa Research Foundation (AwwaRF) is a member-supported, international, nonprofit organization that sponsors research to enable water utilities, public health agencies, and other professionals to provide safe and affordable drinking water to consumers.

The Foundation's mission is to advance the science of water to improve the quality of life. To achieve this mission, the Foundation sponsors studies on all aspects of drinking water, including supply and resources, treatment, monitoring and analysis, distribution, management, and health effects. Funding for research is provided primarily by subscription payments from approximately 1,000 utilities, consulting firms, and manufacturers in North America and abroad. Additional funding comes from collaborative partnerships with other national and international organizations, allowing for resources to be leveraged, expertise to be shared, and broad-based knowledge to be developed and disseminated. Government funding serves as a third source of research dollars.

From its headquarters in Denver, Colorado, the Foundation's staff directs and supports the efforts of more than 800 volunteers who serve on the board of trustees and various committees. These volunteers represent many facets of the water industry, and contribute their expertise to select and monitor research studies that benefit the entire drinking water community.

The results of research are disseminated through a number of channels, including reports, the Web site, conferences, and periodicals.

For subscribers, the Foundation serves as a cooperative program in which water suppliers unite to pool their resources. By applying Foundation research findings, these water suppliers can save substantial costs and stay on the leading edge of drinking water science and technology. Since its inception, AwwaRF has supplied the water community with more than $300 million in applied research.

More information about the Foundation and how to become a subscriber is available on the Web at **www.awwarf.org**.

Occurrence Survey of Boron and Hexavalent Chromium

Prepared by:
Michelle M. Frey and **Chad Seidel**
McGuire Environmental Consultants, Inc., Denver, Colorado

Marc Edwards and **Jeffrey L. Parks**
Virginia Polytechnic Institute and State University, Blacksburg, Virginia

and

Laurie McNeill
Utah State University, Logan, Utah

Sponsored by:
Awwa Research Foundation
6666 West Quincy Avenue, Denver, CO 80235-3098

Published by:

Awwa Research Foundation
Advancing the Science of Water®

IWA Publishing

DISCLAIMER

This study was funded by the Awwa Research Foundation (AwwaRF). AwwaRF assumes no responsibility for the content of the research study reported in this publication or for the opinions or statements of fact expressed in the report. The mention of trade names for commercial products does not represent or imply the approval or endorsement of AwwaRF. This report is presented solely for informational purposes.

Published by IWA Publishing, Alliance House, 12 Caxton Street, London SW1H 0QS, UK

Telephone: +44 (0) 20 7654 5500; Fax: +44 (0) 20 7654 5555; Email: publications@iwap.co.uk;
Web: **www.iwapublishing.com**

AwwaRF report number 91044F. Originally published by AwwaRF for its subscribers in 2004.
IWA Publishing version published 2005.

ISBN 978-1-84339-926-1

CONTENTS

TABLES

FIGURES

FOREWORD

The Awwa Research Foundation is a nonprofit corporation that is dedicated to the implementation of a research effort to help utilities respond to regulatory requirements and traditional high-priority concerns of the industry. The research agenda is developed through a process of consultation with subscribers and drinking water professionals. Under the umbrella of a Strategic Research Plan, the Research Advisory Council prioritizes the suggested projects based upon current and future needs, applicability, and past work; the recommendations are forwarded to the Board of Trustees for final selection. The foundation also sponsors research projects through the unsolicited proposal process; the Collaborative Research, Research Applications, and Tailored Collaboration programs; and various joint research efforts with organizations such as the U.S. Environmental Protection Agency, the U.S. Bureau of Reclamation, and the Association of California Water Agencies.

This publication is a result of one of these sponsored studies, and it is hoped that its findings will be applied in communities throughout the world. The following report serves not only as a means of communication of the results of the water industry's centralized research program but also as a tool to enlist the further support of nonmember utilities and individuals.

Projects are managed closely from their inception to the final report by the foundation's staff and large cadre of volunteers who willingly contribute their time and expertise. The foundation serves a planning and management function and awards contracts to other institutions such as water utilities, universities, and engineering firms. The funding for this research effort comes primarily from the Subscription Program, through which water utilities subscribe to the research program and make an annual payment proportionate to the volume of water they deliver and consultants and manufacturers subscribe based on their annual billings. The program offers a cost-effective and fair method for funding research in the public interest.

A broad spectrum of water supply issues is addressed by the foundation's research agenda: resources, treatment and operations, distribution and storage, water quality and analysis, toxicology, economics, and management. The ultimate purpose of the coordinated effort is to assist water suppliers in providing the highest possible quality of water economically and reliably. The true benefits are realized when the results are implemented at the utility level. The foundation's trustees are pleased to offer this publication as a contribution toward that end.

Walter J. Bishop
Chair, Board of Trustees
Awwa Research Foundation

James F. Manwaring, P.E.
Executive Director
Awwa Research Foundation

ACKNOWLEDGMENTS

This report was made possible by the financial support of the Awwa Research Foundation and the efforts of many people. At the inevitable risk of omitting someone, particular thanks go to the following people for their contributions to the project:

AwwaRF Project Advisory Committee

John Gaston, CH2M Hill
Elizabeth Hedrick, U.S. EPA
Gary Schafran, Old Dominion University
Kenan Ozekin, AwwaRF Project Manager
Albert Ilges, AwwaRF Project Manager

McGuire Environmental Consultants, Inc.

Nicole Graziano
John Rosendahl

Utah State University

Han Lai
Julia Rahman
Brian Thomas
Crystal Viator
Chen Yan

Last, the authors would like to thank the staff of the 189 drinking water utilities that gave of their time and effort to collect water samples for analysis in this project.

EXECUTIVE SUMMARY

Understanding national occurrence patterns for contaminants in drinking water is critical to the development of sound public policy on health protection. Both hexavalent chromium and boron have been proposed to be part of EPA's Contaminant Candidate List (CCL) and are highly probable to have drinking water standards developed within the next five-year cycle of regulatory development. Both of these compounds present complex occurrence and treatment issues for the drinking water community if low-level occurrence is determined to be a public health concern. This project – referred to herein as the National Chromium and Boron Occurrence Survey (NCBOS) – specifically addressed the following questions:

- What are the analytical method challenges and sensitivities for reliable low-level detection of chromium species and boron in drinking water supplies?
- What are the national occurrence patterns for chromium species and boron in drinking water sources?
- What is the fate of these compounds through drinking water facilities and distribution systems?

ANALYTICAL CHALLENGES

This project investigated analytical challenges for both chromium and boron analysis. The investigation focused on interferences, preservation techniques, and digestion methods for accurate and repeatable recovery performance. Laboratory experiments were first used to investigate the analytical challenges. Preliminary field sampling based on knowledge gained from the laboratory experiments validated the analytical methods used for the regular survey.

Interferences in boron analysis were found with aluminum and iron for at least one spectrum evaluated when background concentrations approached 100 mg/L as either Fe or Al. These concentrations are well above those expected in drinking water sources. All samples collected during this survey contained less than 100 mg/L Fe and Al; therefore, these interferences were not problematic. Boron was found to be otherwise unaffected by sample handling and processing methods as long as base preservation methods were employed ($pH > 8$).

Sampling for chromium necessitated consideration of both total chromium and hexavalent chromium analyses. Previous field sampling experiences had resulted in total chromium concentrations less than corresponding hexavalent chromium concentrations (Eaton et al. 2001). A series of lab experiments were conducted to determine the best approach for chromium analysis to correctly quantify both hexavalent and total chromium.

Using in-bottle acid digestion procedures similar to Method 3500-Fe B (APHA, AWWA, WEF 1998) – the hydroxylamine digestion to completely dissolve iron– was found to perform best in achieving complete recovery of total chromium. Base preservation for hexavalent chromium was found to be adequate when the sample pH was raised above pH 8.

The final survey sample kit preservation, digestion, and sample analysis procedures were developed based on results from the investigation of both chromium and boron analytical challenges. The final survey sample kit included two bottles: one (Bottle A) acid-preserved and the other (Bottle B) base-preserved. Sample kits were sent from Virginia Tech to participating drinking water utilities that collected samples following detailed sampling instructions. After samples kits were returned to Virginia Tech by the participating utilities, the acid-preserved sample was digested in the bottle with hydroxylamine following a procedure similar to that specified in Standard Method 3500 – Fe B (APHA, AWWA, WEF 1998). Following this procedure, 4% hydroxylamine and 2% hydrochloric acid were added to the bottle. The bottle was then heated to 85°C for 24 hours. The resulting digested solution was then filtered using a 0.45 μm pore size filter and analyzed for total boron by ICP-ES following Standard Method 3120 B (APHA, AWWA, WEF 1998) at Virginia Tech. An aliquot of the hydroxylamine digested and filtered sample was sent to Utah State University (USU) for total chromium analysis by ICP-MS following EPA Method 200.8 (Determination of Trace Elements in Waters and Wastes by Inductively Coupled Plasma – Mass Spectrometry) (USEPA 1994). The base-preserved sample was filtered in the Virginia Tech lab and then sent to USU for hexavalent chromium analysis by IC following EPA Method 1636 (Determination of Hexavalent Chromium by Ion Chromatography) (USEPA 1996a). Figure ES-1 describes the sample kit preservation and subsequent digestion, filtration, and sample analysis procedures used for the regular occurrence survey samples.

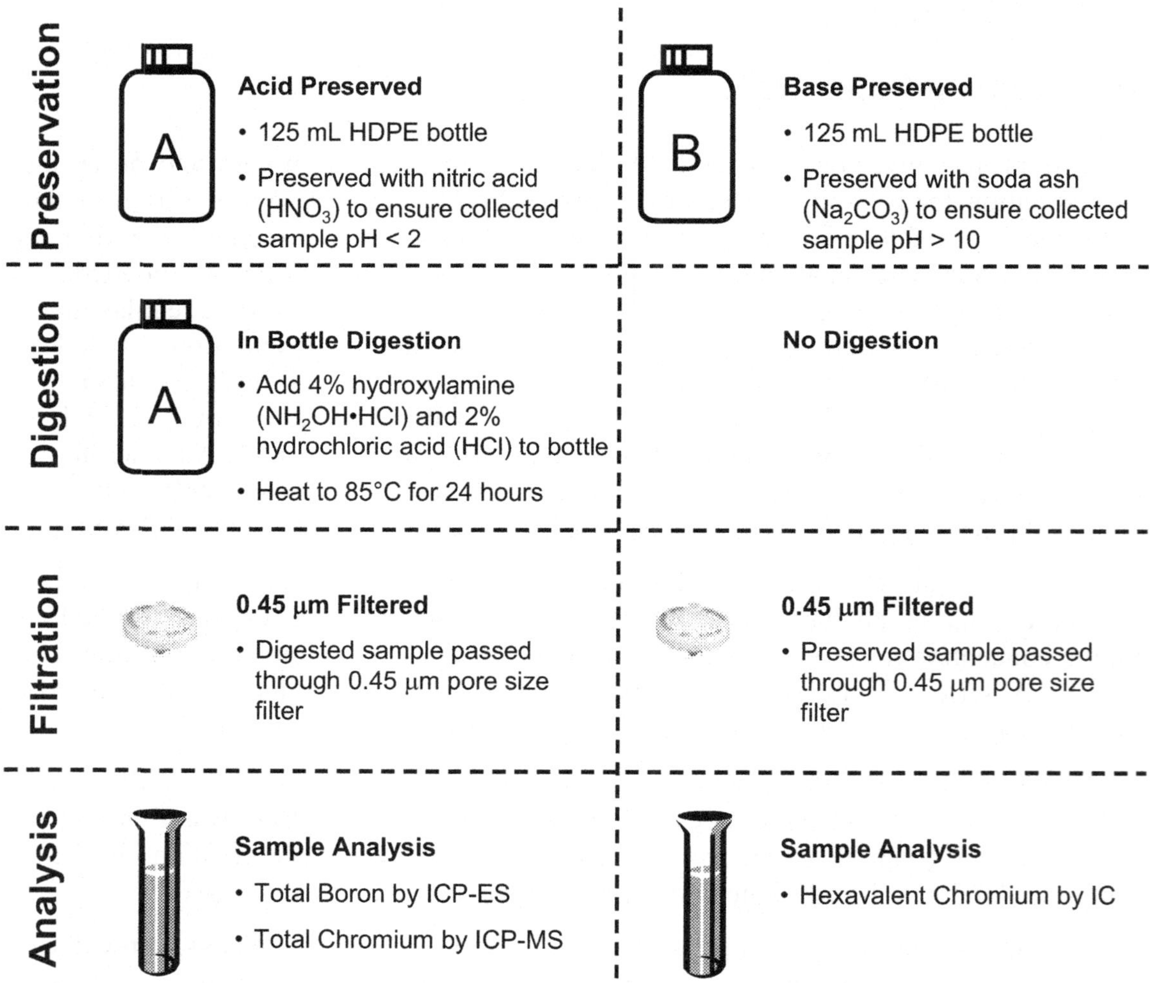

Figure ES-1. Sample collection and subsequent digestion and analysis used for the regular occurrence survey samples

NATIONAL OCCURRENCE TRENDS

The project team recruited 189 participating utilities representing 407 source waters. The participating utilities are distributed across 41 states, with the greatest number of utilities located in California, followed by Illinois and Indiana. Of the 407 source waters recruited to be sampled, 273 (67%) were groundwater sources and 134 (33%) were surface water sources. Of the 407 source water sample kits sent to participating utilities in both the preliminary and regular surveys, 342 sample kits were returned and analyzed.

The occurrence results from the NCBOS for all source water samples are summarized in Table ES-1.

Table ES-1
NCBOS occurrence data summary

Parameter	Count	Average (µg/L)	Median (µg/L)	Minimum (µg/L)	Maximum (µg/L)
Boron	341	167.9	40.0	< MDL	3,320
Total Chromium	342	2.0	0.8	< MDL	47.1
Hexavalent Chromium	341	1.1	< MDL	< MDL	52.6

Boron MDL = 2.0 µg/L
Total Chromium MDL = 0.6 µg/L
Hexavalent Chromium MDL = 0.2 µg/L

Several important results were identified from the NCBOS data. Boron was found to occur higher in groundwater sources than surface water sources, but it occurred equally across water system size categories. Source waters treated by NF/RO membranes exhibited the highest boron occurrence by treatment type, followed by MF/UF, softening, and disinfection only. The NF/RO membrane treated source waters were typically brackish which contain higher boron concentrations than typical fresh waters. No clear occurrence patterns exist by EPA region for groundwater boron occurrence; however, surface water boron occurs to a greater degree in the western U.S than in the eastern U.S.

A majority of the hexavalent chromium sample results were found to be less than the MDL of 0.2 µg/L. Total chromium was found to occur equally in surface waters and groundwaters; however, hexavalent chromium is not found in surface waters to nearly the same degree as in groundwaters. NCBOS total chromium concentrations in surface waters are – with few exceptions – composed primarily of trivalent chromium. Several groundwater results show total chromium concentrations composed exclusively by hexavalent chromium. This speciation trend by source type, which has not been previously reported, is illustrated in Figure ES-2 where paired total and hexavalent chromium concentration are plotted for both groundwaters and surface waters.

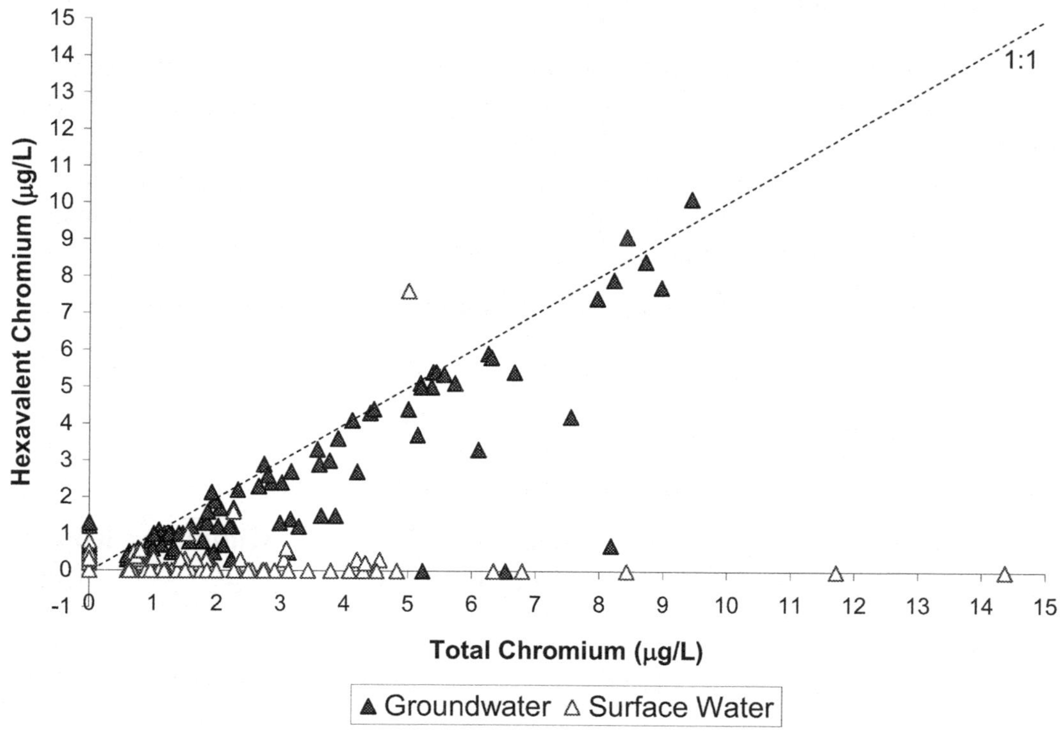

Figure ES-2. Paired total and hexavalent chromium results (below 15 µg/L) by source type

When evaluating all source waters, both total and hexavalent chromium exhibited elevated concentrations in the medium size category (10,001 – 50,000 population served) – a size category represented primarily by ground waters in this survey which have higher chromium concentrations.

The occurrence trends for total and hexavalent chromium are essentially the same for all treatment types except conventional treatment where total chromium concentrations are higher. Since these treatment types are typically associated with surface waters, this result is consistent with the chromium speciation occurrence trends by source water previously reported. This conclusion can also be made for waters treated by softening. The highest chromium concentrations are found in source waters not treated or just disinfected – typically associated with groundwaters.

When evaluating regional occurrence trends, chromium – both total and hexavalent – is greatest in EPA Region 9 which includes California. Total chromium occurrence is more variable by region in surface waters than groundwaters. No clear regional patterns exist for total chromium in surface waters. However, regional patterns did exist for groundwater results where total chromium occurred to the greatest degree in EPA Region 9.

TREATMENT PROFILE SURVEY FOR CHROMIUM AND BORON

The fate of chromium and boron in water systems was investigated by performing a special survey where longitudinal sampling was performed from source-to-tap in fifteen water systems.

The results found from the treatment profile survey for hexavalent and total chromium indicated the following:

- Temporal variability in chromium concentrations should be expected for both surface and ground water supplies. This variation may include chromium speciation behavior.
- Hexavalent chromium was the dominant form of chromium in the potable water sources and treated supplies sampled during the treatment profile survey (Table ES-2). However, the effects of oxidants on chromium speciation could not be distinguished from the treatment profile survey results. No reductants were used by the systems profiled.
- In the case of conventional treatment systems (coagulation with filtration), trivalent chromium appears to be effectively removed while hexavalent chromium is not removed.
- Other than the ability of conventional treatment to remove trivalent chromium, no other technology included in the profile survey affected hexavalent chromium or total chromium levels.

Table ES-2
Chromium speciation in treatment profile utilities

Source Type	Treatment Type[1]	Disinfection Type	Raw Water Levels, µg/L		Treated Water Levels, µg/L		Distribution System Levels, µg/L		Cr(VI) to Total Cr Ratio, %[2]		
			Cr(VI)	Tot Cr	Cr(VI)	Tot Cr	Cr(VI)	Tot Cr	Raw	Treated	Dist System
Ground	**Disinfection Only ****	**CL2**	**22.2**	**31.2**	**Not Sampled**		**0.4**	**< 0.1**	**71%**	**NS**	**> 100%**
Ground	Disinfection Only	CL2	0.7	1.4	0.8	0.6	0.7	0.6	51%	135%	106%
Ground	**Disinfection Only ****	**CL2**	**14.0**	**13.2**	**13.8**	**12.9**	**8.4**	**6.6**	**106%**	**107%**	**127%**
Ground	Disinfection Only	CL2	1.3	2.2	4.2	4.2	11.9	11.4	62%	99%	104%
Ground	Fe/Mn	CL2	< 0.2	0.4	< 0.2	0.1	< 0.2	2.8	0%	0%	0%
Ground	Fe/Mn	CL2/CLM	< 0.2	< 0.1	< 0.2	< 0.1	0.4	0.4	I	I	95%
Ground	**Fe/Mn ****	**CL2/CLM**	**< 0.2**	**0.1**	**< 0.2**	**< 0.1**	**< 0.2**	**0.3**	**0%**	**I**	**0%**
Ground	**GAC ****	**CL2/CLM**	**11.3**	**9.3**	**10.8**	**10.2**	**< 0.2**	**0.2**	**122%**	**106%**	**0%**
Ground	**GAC ****	**CL2**	**4.9**	**5.0**	**5.5**	**4.8**	**3.4**	**3.3**	**98%**	**114%**	**103%**
Ground	NF/RO	CL2/CLM	0.6	0.3	< 0.2	< 0.1	Not Sampled		200%	I	NS
Ground	**None ****	**NONE**	**17.8**	**16.6**	**Not Sampled**		**17.6**	**16.2**	**107%**	**NS**	**109%**
Surface	**Alum ****	**CL2**	**< 0.2**	**6.1**	**< 0.2**	**< 0.1**	**< 0.2**	**< 0.1**	**0%**	**I**	**I**
Surface	**Alum ****	**CL2**	**1.8**	**2.3**	**1.5**	**1.3**	**1.7**	**1.6**	**80%**	**117%**	**109%**
Surface	Alum	CL2	< 0.2	0.4	< 0.2	0.2	< 0.2	0.3	0%	0%	0%
Surface	Alum	CL2/CLM	< 0.2	0.4	< 0.2	< 0.1	0.2	0.4	0%	I	55%
Surface	Ferric	O3/CL2/CLM	< 0.2	0.3	< 0.2	0.1	< 0.2	0.1	0%	0%	0%

1. Those systems included in the treatment profile study to specifically capture the behavior of chromium are indicated by '**' and these results are shown in **BOLD**.
2. As appropriate to the treatment system, chromium speciation fractions were not determined when either (a) a location was not sampled because the intermediate location was not available to the system type (NS), or (b) the analytical results for both Cr(VI) and total chromium were below detection and the calculation of a speciation fraction was indeterminate (I).

Boron was not effectively removed through the treatment systems investigated in the treatment profile survey (Figure ES-3). Even removal by the NF/RO groundwater system was negligible (15%). The maximum apparent removals ranged from 25% to 30% and occurred in a conventional surface water (alum) system and a groundwater system where iron/manganese treatment was practiced. These calculated removals are likely artifacts due to combinations of blending and analytical variability within the plants and samples collected. Changes in boron concentration between the treatment process and the distribution system were also found to generally be negligible; exceptions may be explained by blending with different source waters.

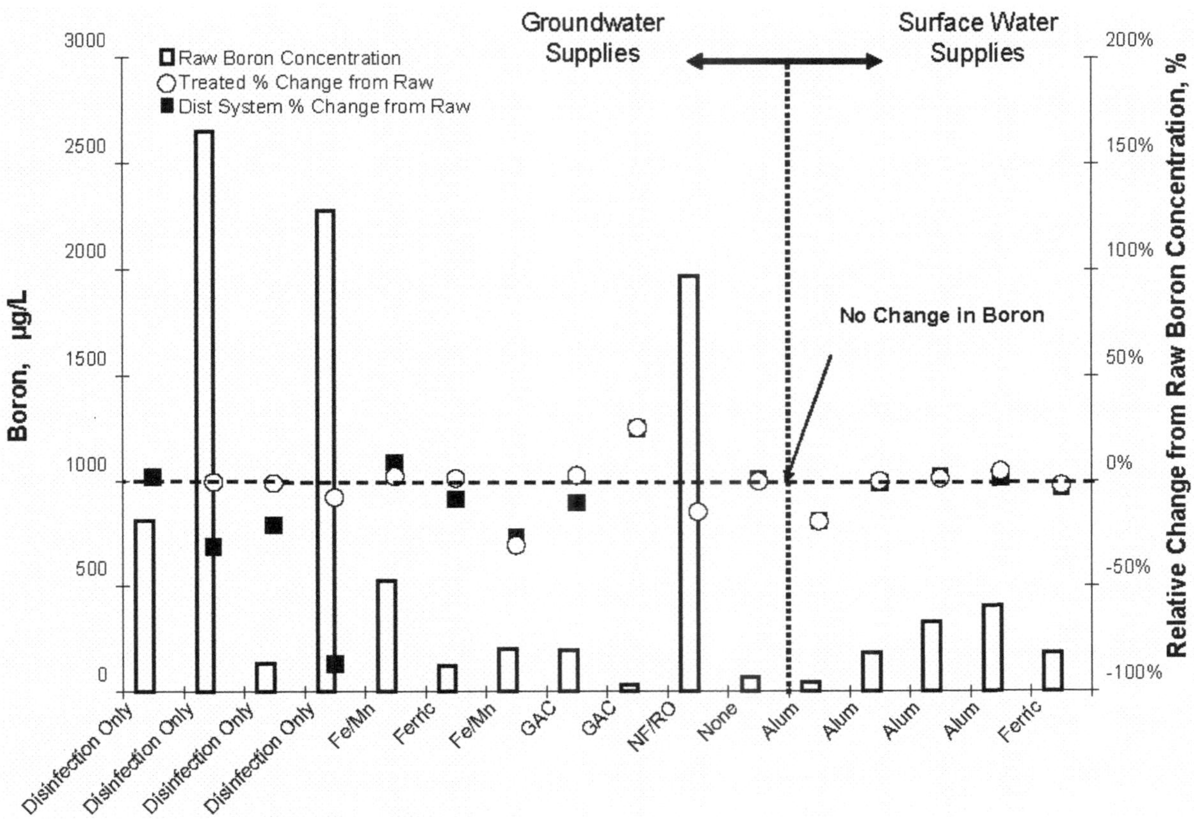

Figure ES-3. Treatment profile results for boron.

Based on these results, boron levels in source waters would be expected to remain fairly conservative through most water systems. Other technologies with potential to remove boron concentrations to a greater degree than those observed in this study may include lime softening and high pH membrane technologies. With few exceptions, source water boron represents a reasonable estimate of boron levels delivered to customers' taps.

CHAPTER 1
UNDERSTANDING CHROMIUM AND BORON ISSUES FOR DRINKING WATER

Understanding national occurrence patterns for contaminants in drinking water is critical to the development of sound public policy on health protection. The roles of hexavalent chromium (Cr(VI)) and boron (B) in the public policy debates for drinking water are still quite ambiguous. The question for water utilities and public health agencies is "what are the public health consequences of low levels of Cr(VI) and boron in drinking water supplies?" Limited evidence of potential adverse health effects of low-level exposure to Cr(VI) led the State of California to initially adopt a Public Health Goal (PHG) of 2.5 μg/L for total chromium which has subsequently been rescinded (OEHHA 1999). This goal was intended to provide a protective level of exposure for Cr(VI) of 0.2 μg/L due to assumptions about the speciation of chromium in water supplies. The two common species of chromium in drinking water have distinctly different public health consequences. Trivalent chromium (Cr(III)) appears to have little to no associated human toxicity – in fact, it is a trace nutrient – while Cr(VI) is classified as a "known human carcinogen" via inhalation exposure and is suspected to cause adverse reproductive and developmental effects. As such, the key to public health protection is control of the Cr(VI) specie.

When the California PHG was established, the State assumed that Cr(VI) accounted for only 7.2% of the total chromium concentration. This was based on very limited occurrence data from one published study (Kacynski and Kieber 1993). Since that time, improved analytical methods have become available and increased monitoring of groundwater in the San Fernando Valley, for example, have shown that Cr(VI) often composes more than 70% of the total chromium present in these supplies (Figure 1-1).

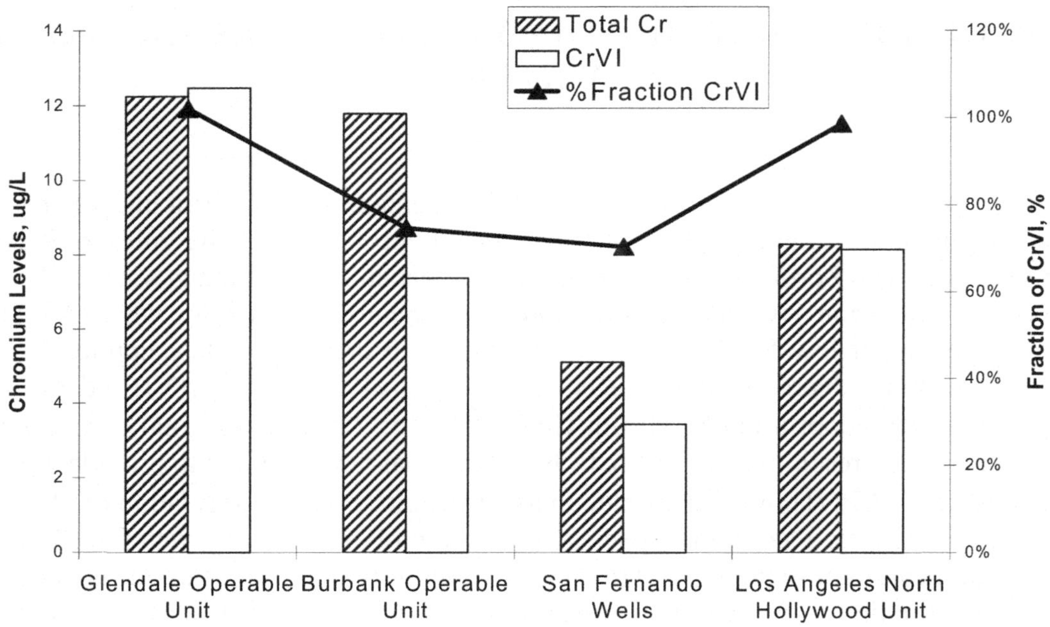

Figure 1-1. Chromium concentrations and speciation in the San Fernando Valley

If this new information on speciation had been applied, the PHG for total chromium would have been lowered to nearly the value of the health protective level for Cr(VI), 0.2 μg/L. In the case that a drinking water standard is set for Cr(VI) – or revised for total chromium with the new information on chromium speciation – two important consequences would result: (1) naturally occurring Cr(VI) in existing water supplies could lead to compliance challenges, and (2) the national implications of such a standard would be profound, impacting a significant portion of the drinking water industry.

Both hexavalent chromium and boron have been proposed to be part of EPA's Contaminant Candidate List (CCL) and are highly probable to have drinking water standards developed within the next five-year regulatory development cycle. Boron occurrence in drinking water – if regulated at levels that impact a substantial portion of the public water supplies in the U.S. – presents a unique challenge to the drinking water industry. The chemistry of boron in water is such that removal is very difficult, and those systems that require treatment for boron are likely to require the application of advanced technologies such as reverse osmosis. It is of interest to the drinking water community that seawater contains substantially higher boron concentrations than most fresh water supplies. As such, desalination facilities will especially be concerned about future boron regulations and the likely performance of reverse osmosis treatment for boron removal.

Both of these compounds present complex occurrence and treatment issues for the drinking water community if low-level occurrence is determined to be a public health concern. This project – referred to herein as the National Chromium and Boron Occurrence Survey (NCBOS) – specifically addressed the following questions:

- What are the analytical method challenges and sensitivities for reliable low-level detection of chromium species and boron in drinking water supplies?

- What are the national occurrence patterns for chromium species and boron in drinking water sources?
- What is the fate of these compounds through drinking water facilities and distribution systems?

CHROMIUM CHEMISTRY

Chromium exhibits a complex chemistry and occurs in various forms in surface and groundwaters. Naturally occurring chromium primarily results from weathering of chromite ($FeO \cdot Cr_2O_3$) and other chromium-bearing minerals present in bedrock and soils. Chromium is the 21st most abundant element in the earth crust (Nriagu 1988). Anthropogenic sources include discharges from tanneries, metal dipping, electroplating and industrial process waters to which chromium is added for corrosion control.

Chromium, a metallic element with an atomic number of 24, is a member of periodic table group VIB along with molybdenum and tungsten. Chromium has four naturally occurring isotopes, but none of them is radioactive (Weast, Astle, and Beyer 1988). Chromium forms a number of salts, which are characterized by a variety of colors, solubility values, and other properties. The most important chromium salts are sodium and potassium chromates and dichromates, plus the potassium and ammonium chrome alums (Hodgman, Weast, and Selby 1961). Chromium has several oxidation states (Figure 1-2A), the most common and stable of which are Cr(II), Cr(III) and Cr(VI) (Baes and Mesmer 1976). Several Cr(IV) and Cr(V) species are known as intermediates in redox reactions and are unstable with respect to Cr (III) and Cr(VI).

In aquatic systems most chromium occurs in two oxidation states, Cr(III) and Cr(VI). Cr(III) occurs as a cation, and the hydroxide complex is insoluble. Cr(VI) occurs as an anion as either chromate ($HCrO_4^-/CrO_4^{2-}$) or dichromate ($Cr_2O_7^{2-}$). Both anionic forms are very soluble, and the formation of each is pH-dependent (Sengupta, Clifford, and Subramonian 1986). The simple ionic form of Cr(III) is $Cr(III)^+$, which predominates at pH <4. At pH >4, $Cr(III)^+$ forms hydroxide complexes in a stepwise fashion as pH increases ($Cr(III)^+ \rightarrow Cr(OH)^{2+} \rightarrow Cr(OH)_2^+ \rightarrow Cr(OH)_3^0 \rightarrow Cr(OH)_4^-$). These complexation reactions control the ionic state of aqueous Cr(III), with the ionic charge changing from +3 to –1 between pH 4 and pH 10. At a pH range of 6–8, which is typical for natural water supplies, the predominant aqueous species is $Cr(OH)_3^0$. However, at this pH range Cr(III) exhibits low solubility, which is controlled by $Cr(OH)_3(s)$ (K_{SP} of $\approx 10^{-30}$). The minimum Cr(III) solubility is at pH ≈8.

In contrast to Cr(III), Cr(VI) is highly soluble. At low concentrations, Cr(VI) is present in water as diprotic chromic acid (H_2CrO_4, $pK_{a1} = 0.81$, $pK_{a2} = 6.49$) (Butler 1967, Tong and King 1953). Thus, in natural water supplies two Cr(VI) oxyanion species predominate, monovalent $HCrO_4^-$ below pH 6.5 and divalent CrO_4^{2-} above pH 6.5 (Figure 1-3). An additional Cr(VI) species, dichromate ($Cr_2O_7^{2-}$), predominates at concentrations greater than 1 g/L. It is unlikely that any drinking water source would contain such a high Cr(VI) concentration. However, it is possible that Cr(VI) concentrations at the surface of a treatment media may be high enough to favor the presence of dichromate.

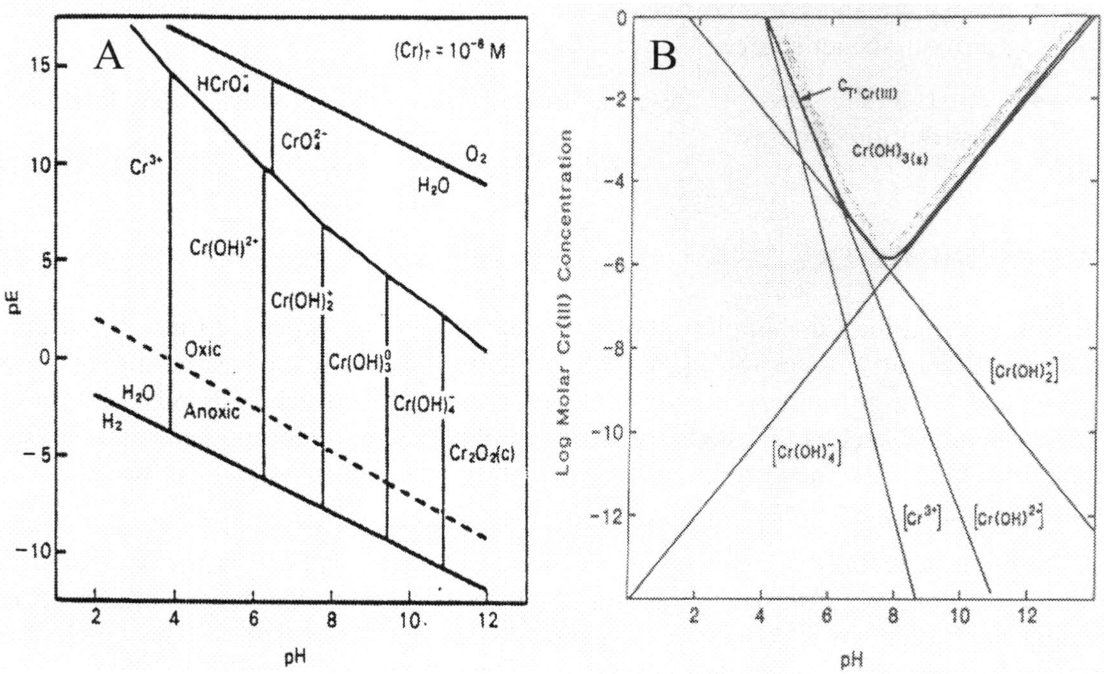

Figure 1-2. (A) pE–pH diagram of aqueous chromium; (B) Cr(III) solubility at 25°C

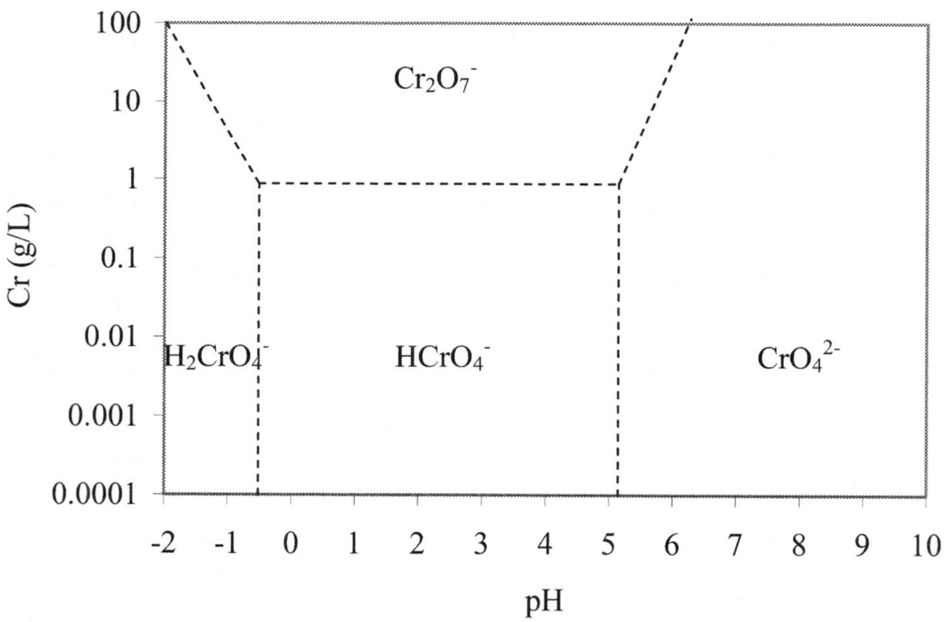

Figure 1-3. Relative distribution of Cr(VI) species in water as a function of pH and Cr(VI) concentration

CHROMIUM OCCURRENCE

Detailed studies of chromium occurrence and speciation in natural waters are lacking. Durfor and Becker (1964) report a median value of 0.43 μg/L total chromium in drinking source waters, while Kharkar et al.(1968) report an average total chromium value in river water of 1.4 μg/L. However, both of these studies are dated and do not address speciation. A hexavalent chromium sampling program was initiated by the State of California Department of Health Services (CalDHS) in 2000 to evaluate statewide occurrence. As of Dec. 1, 2003, 33% of the 6,229 drinking water sources monitored contained detectable concentrations of Cr(VI). Table 1-1 summarizes the DHS findings.

Table 1-1
Detections of hexavalent chromium in California drinking water sources*†

Peak concentration (μg/L)	Number of sources	Percentage of sources %
>50	5	-
46–50	2	-
41–45	3	-
36–40	5	-
31–35	7	-
26–30	20	-
21–25	22	-
16–20	59	1
11–15	134	2
6–10	406	7
1–5	1,346	22
Total Cr(VI) detections	2,068	33
Sources with no data or <1 μg/L	4,161	67
Total sources reporting	6,229	100

* As of Dec. 1, 2003

† Cr(VI) concentrations are from sources reporting more than a single detection. Sources may include both untreated and treated supplies, distribution systems, blending reservoirs, and other sampled entities. Table does not include agricultural wells, monitoring wells, or more than one representation of the same source (e.g., a source with data for both untreated and treated supplies is counted as a single source). Detections lower than the detection limit for reporting purposes are not included. Data should be considered draft.

BORON CHEMISTRY

Elemental boron does not exist in nature, but is always found combined with oxygen in the form of inorganic borates or boric acid. Boron is a non-metallic, electron deficient element that forms compounds by reacting with electronegative elements such as oxygen.

In nature, boron is found in the form of boric acid, borate (i.e. salt of boric acid), or as a borosilicate mineral (Holleman and Wiberg, 2001). Boric acid, H_3BO_3 (or $B(OH)_3$), behaves as a weak Lewis acid in aqueous solution (Power and Woods, 1997). It accepts hydroxide ion from water and releases a proton into solution according to the following equilibrium equation ($K_a = 5.8x10^{-10}$; $pK_a = 9.24$ @ 25°C) (Dean, 1987):

$$B(OH)_3 + H_2O <=> B(OH)_4^- + H^+$$

Boric acid dissociates as a function of pH; above pH 9.24 the anion, $B(OH)_4^-$, is predominant, while below pH 9.24 the uncharged species is predominant. Boric acid is soluble in water (5.5 g / 100 g solution at 25 °C) and its solubility increases with temperature (Waggott, 1969). At concentrations below 0.02 M (216 mg/L as B) only the mononuclear species B(OH)3 and B(OH)4- are present. Boric acid is soluble in water (5.5 g / 100 g solution at 25 °C) and its solubility increases with temperature (Waggott, 1969). At concentrations below 0.02 M (216 mg/L as B) only the mononuclear species $B(OH)_3$ and $B(OH)_4^-$ are present.

Boron is known to form soluble complexes with organic polyhydroxy compounds, either present naturally or added to water specifically to complex boron (Mackin, 1986; Smith et al, 1995), and a host of bio-molecules in mammalian tissues and fluids (Woods, 1996). Complexation can increase the acidity of boric acid which acts as a Lewis Acid. For example, mannitol complexation produces a negatively charged complex with greatly increased acidity. With respect to organic matter, the extent of natural boron complexation is believed to be about 20% or less in natural waters (Mackin, 1986), and boron is released during organic matter oxidation which occasionally leads to inverse relationships between dissolved oxygen and dissolved boron in the water column (Shirodkar et al., 1992). Boron also forms soluble complexes with fluoride ion at lower pH (e.g., BF_4^-),

There is little or no information in the environmental literature on complexes between boron at the mg/L level and environmentally important cations such as Al^{+3}, Fe^{+3}, Fe^{+2}, Ca^{+2}, Mg^{+2}, K^+, Na^+, Cu^+, and Cu^{+2}. Likewise no solubility product values could be readily found for solids containing these metals, although it has been noted that the solubility of calcium borate is approximately 600 mg/L as boron (Lapp et al., 1976). It is generally believed that precipitates are only suited to removing boron to the 10-100 mg/L range.

BORON OCCURRENCE

Various surveys have been performed to determine the distribution of boron in United States waters. In 1969 the United States Public Health Service conducted a survey of 969 community water systems and found a 99th percentile boron concentration of 1.0 mg/L; i.e., 99% of all waters tested had boron concentrations less than 1.0 mg/L. The maximum level encountered was 3.28 mg/L (Coughlin, 1998). In 1987 the National Inorganics and Radionuclides Survey reported a mean boron concentration of 0.15 mg/L for 989 public water supplies tested. The maximum level encountered in this survey was 4.0 mg/L (Coughlin, 1998).

CHAPTER 2
ANALYTICAL METHOD PERFORMANCE FOR CHROMIUM AND BORON

This project began with preliminary efforts to determine the most effective method to collect, preserve, and handle survey samples. Special attention was given to material selection, quantity and type of preservative, method detection limits, constituent recovery, and potential interferences. A series of laboratory bench-level tests were conducted at Virginia Tech to confirm the performance of the analytical methods. Following the laboratory analytical methods investigation, a preliminary survey was performed to compare the analytical methods in a broad array of natural water matrices. This chapter discusses the results of the analytical method performance investigations and presents the final survey sample kit preservation, digestion procedures, and sample analysis procedures.

In addition to the analytical methods investigation described in this chapter, a significant effort was undertaken outside the scope this project specifically addressing the determination of total chromium in environmental water samples (Parks et al. 2004). It should be noted that the analytical methods ultimately used for chromium analysis in this survey rely upon the detailed assessment provided by Parks et al.

SAMPLE KIT DEVELOPMENT

An evaluation of the materials and methods for field sampling included selection of preservatives, sampling kit components, and reagent water.

Preservative

Sodium carbonate (Na_2CO_3) was chosen as the preservative for Cr(VI) and boron field sampling. A pH above 8 is required for chromium speciation, while a pH above 10 is required for boron preservation (Downing et al. 1998). Therefore, the first step in developing a robust sampling kit for the survey was to determine an appropriate soda ash dose to raise the sample pH above 10 for virtually any source water to be sampled. A hypothetical source water with an high alkalinity of 500 mg/L as $CaCO_3$ and a pH of 7.5 was considered as a worst case condition that could be encountered in the survey. Based on this worse case condition, the dose of 12 mM soda ash required to raise pH above 10 was determined.

Following EPA protocols, acid preservation by nitric acid was used for total chromium and boron samples. The amount of acid required was based on lowering the pH of any sample to below 2. The required dose of concentrated nitric acid was determined using the high alkalinity water quality as noted above.

Field Sampling Kit Components: Containers, Filters, Syringes

High-density polyethylene (HDPE) bottles were tested to determine if boron or chromium leached into solution from the container wall or was sorbed from solution to the container wall. In the same series of tests, plastic syringes and 0.45 µm pore size nylon syringe filters were also evaluated.

Eighteen (18) new 250 mL HDPE bottles were filled with nanopure water and 12 mM soda ash. Concentrated nitric acid was added to half the bottles (2% by volume final concentration), resulting in half of the bottles with pH less than 2 and the other half with pH above 10. Boron (40 µg/L) and Cr(VI) (20 µg/L) were added to six (6) bottles while boron (40 µg/L) and Cr(III) (20 µg/L) were added to six (6) other bottles. The remaining six (6) bottles were 'blanks' and had no boron or chromium added. Samples were withdrawn and filtered using the same syringes and filters used during the regular survey at 1 hour, 6 days, and 20 days from each of the eighteen bottles. Samples were analyzed at Virginia Tech for boron and total chromium using an inductively coupled plasma (JY Ultima ICP-ES). All bottles were then shipped to Utah State University (USU) for chromium analysis utilizing ion chromatography for Cr(VI) (Dionex DX-320) and/or graphite furnace for total chromium (Perkin Elmer AAnalyst). Samples were withdrawn, filtered, and analyzed at USU 88 days after the initial chromium spike. Table 2-1 contains the chromium results from these tests. In all cases, agreement was found between the two sets of independent analyses performed at Virginia Tech and USU for Cr(VI) and total chromium.

The results indicate that neither boron or Cr(VI) leached from or sorbed to the HDPE sample bottles. However, up to 47% of the Cr(III) was lost – potentially by sorption or precipitation – in the bottles with soda ash preservative but no acid (i.e. High pH > 10 No significant Cr(III) was lost in acid-preserved samples. These observations indicate that Cr(III) would not be recovered from soda ash preserved samples during the survey unless acid digestions are subsequently performed in the actual sample bottles.

Table 2-1
Evaluation of chromium fate in sample container materials

Water Matrix Condition	Chromium Results (μg/L) at Sample Holding Time, days			
	1	6	20	88*
Blank, High pH				
Aliquot 1	3.0	7.0	1.3	-0.1
Aliquot 2	2.6	3.2	0.9	0
Aliquot 3	2.2	6.1	0.7	0
Average	2.6	5.4	1.0	0
Standard Deviation	0.4	2.0	0.3	0
Blank, Low pH				
Aliquot 1	0	-0.4	0.4	0
Aliquot 2	0	-0.6	-0.3	0
Aliquot 3	-0.4	-4.7	0.1	0.1
Average	-0.1	-1.9	0.1	0
Standard Deviation	0.2	2.4	0.4	0.1
20 μg/L Cr(VI), High pH				
Aliquot 1	22.7	25.9	21.5	19.9
Aliquot 2	22.4	25.7	21.1	20.1
Aliquot 3	22.7	25.0	21.5	20.3
Average	22.6	25.5	21.4	20.1
Standard Deviation	0.2	0.5	0.2	0.2
20 μg/L Cr(VI), Low pH				
Aliquot 1	20.5	21.0	19.8	20.4
Aliquot 2	20.7	22.0	20.0	20.9
Aliquot 3	20.1	21.2	19.3	20.8
Average	20.4	21.4	19.7	20.7
Standard Deviation	0.3	0.5	0.4	0.3
20 μg/L Cr(III), High pH				
Aliquot 1	21.5	15.3	13.6	16.6
Aliquot 2	21.4	15.6	10.0	17.2
Aliquot 3	22.3	18.0	10.9	12.0
Average	21.7	16.3	11.5	15.3
Standard Deviation	0.5	1.5	1.9	2.8
20 μg/L Cr(III), Low pH				
Aliquot 1	21.0	21.5	20.7	20.5
Aliquot 2	20.2	19.7	21.3	20.9
Aliquot 3	20.5	19.3	20.1	20.6
Average	20.6	20.1	20.7	20.7
Standard Deviation	0.4	1.2	0.6	0.2

* 88-day holding time samples were analyzed by USU laboratory while all other sample results represent analyses by Virginia Tech laboratory.

Initial testing indicated a serious problem with Cr(III) and Cr(VI) sorption to nylon syringe filters. After extensive consideration of the problem and numerous calls to the manufacturer, it was determined that a change in suppliers had occurred, and that a different membrane was being used on the filter. The manufacturer subsequently changed back to the original filter supplier and filter membrane. The manufacturer did not provide information about difference in filter membrane material or manufacturing. A second round of tests was performed and confirmed that no boron or Cr(VI) leached from the syringes or filters. Additional tests were performed to evaluate the sorption of boron or chromium to the nylon syringe filters. Results of these tests indicated no boron sorption to the nylon filter at any pH between 5 and 12.9. Cr(III) did not sorb to filters at pH 11 or below. However, a large amount of Cr(III) sorbed at conditions equivalent to those found in the alkaline digestion procedure (i.e. pH > 12.5). This chromium could be recovered from the filter by subsequently rinsing with a 2% acid solution. These results imply Cr(III) precipitates at high pH; this precipitate is removed by a filter, but can be dissolved fairly quickly in acid.

Reagent Water

Boron-specific ion exchange resin was obtained from Dow Chemical (Product # XUS-43594.00) to remove boron if it was present in reagent grade water. Soda ash (12 mM) was added to samples of tap water, distilled water, nanopure water, and resin-treated nanopure water and each of the waters was then analyzed for boron (Table 2-2).

Table 2-2
Boron levels in reagent water

Water Matrix	Mean Boron Level, µg/L	Standard Deviation of Boron Levels, µg/L
Tap Water	8.38	0.41
Distilled Water	BDL	NA
Reagent Grade Water	BDL	NA
Resin-Treated Reagent Water	BDL	NA

BDL = Below detection limit of 2.0 µg/L
NA = Not applicable

There is no significant statistical difference between Virginia Tech reagent grade and resin-treated water concentrations and in fact both are below the method detection limit (MDL). Based on these results we determined that the reagent grade water was sufficient without the additional ion exchange treatment, but this will be checked during each run of the occurrence study using resin treated water.

EVALUATION OF SAMPLE ANALYSIS INTERFERENCES

Nitrate, iron, and aluminum were evaluated for their interference with boron and chromium sample analysis

Nitrate

Initially, some concern was raised that nitrate may interfere with the boron analysis based upon the experience of Goulden and Kakar (1976). These authors used the curcumin method for boron determination and found that nitrate at levels up to 700 mg/L did not interfere with boron measurements (Goulden and Kakar, 1976). An ICP-ES was used for boron measurements and interference from nitrate was not expected at levels normally found in drinking water sources. As a check, calibration standards were prepared with and without 2% nitric acid (equivalent to 20,700 mg/L). No appreciable difference was detected upon ICP-ES analysis.

Iron

Iron could affect the boron and chromium analysis by either positively or negatively interfering with the analytical method. To test whether an interference existed in the analytical method, three calibration standards were prepared containing 200 μg/L boron, 200 μg/L Cr(VI), and 1, 10, or 100 mg/L of iron. Multiple sample analysis wavelengths for both boron and chromium were analyzed to determine if a specific wavelength was more susceptible to iron interference. Results shown in Table 2-3 indicate that iron is a significant positive interference for boron (wavelength = 249.77 nm) at a level of 100 mg/L. This iron concentration is well above that expected in drinking water sources. In fact, all samples collected during this survey contained less than 100 mg/L Fe and this interference was not problematic.

Table 2-3
Iron interference results for boron and chromium

Sample Description			Boron Results (μg/L) by Spectrum Wavelength			Chromium Results (μg/L) by Spectrum Wavelength		
B (μg/L)	Cr(VI) (μg/L)	Fe (mg/L)	208.96	249.68	249.77	206.15	267.72	284.33
200	200	1	193.1	194.8	193.6	196.5	197.3	192.4
200	200	10	206.2	205.6	211.5	196.8	200.4	196.6
200	200	100	206.7	211.4	322.5	199.7	193.1	206.0

Aluminum

Aluminum interference with boron and chromium sample analyses were also evaluated following the same procedure used to evaluate iron interferences. Results shown in Table 2-4 indicate that aluminum is a positive interference for boron (wavelength = 208.96 nm) at a level of 100 mg/L. Aluminum also has a slight positive interference for chromium (wavelength = 206.15 nm) at a level of 100 mg/L. This aluminum concentration is well above that expected in drinking water sources. In fact, all samples collected during this survey contained less than 100 mg/L Al and this interference was not problematic.

Table 2-4
Aluminum interference results for boron and chromium

Sample Description			Boron Results (μg/L) by Spectrum Wavelength			Chromium Results (μg/L) by Spectrum Wavelength		
B (μg/L)	Cr(VI) (μg/L)	Al (mg/L)	208.96	249.68	249.77	206.15	267.72	284.33
200	200	1	196.1	193.7	192.3	214.8	207.7	209.1
200	200	10	208.8	205.9	206.6	209.6	204.4	204.8
200	200	100	243.9	213.5	212.3	229.1	200.4	204.4

EVALUATION OF IRON SOLIDS ON BORON AND CHROMIUM RECOVERY

It is possible that iron particulate matter might affect boron or chromium concentrations measured in solution. Soluble boron or chromium present in raw water can sorb to iron hydroxide solids or become incorporated into the iron hydroxide crystalline structure. Since many drinking water sources contain appreciable iron concentrations, lab experiments were conducted to evaluate boron and chromium recovery in the presence of iron particles.

Three fractions can be defined in association with an iron hydroxide solid:

1. the "Soluble" fraction passes through a 0.45 μm pore size filter,
2. the "Sorbed" fraction is chemisorbed to the iron hydroxide solid, but can be released in either acidic or basic solution without complete dissolution of the solid,
3. the "Fixed" fraction is associated with the iron hydroxide solid and is not released unless the solid is completely dissolved.

Figure 2-1 illustrates these three fractions using Cr(VI) as an example.

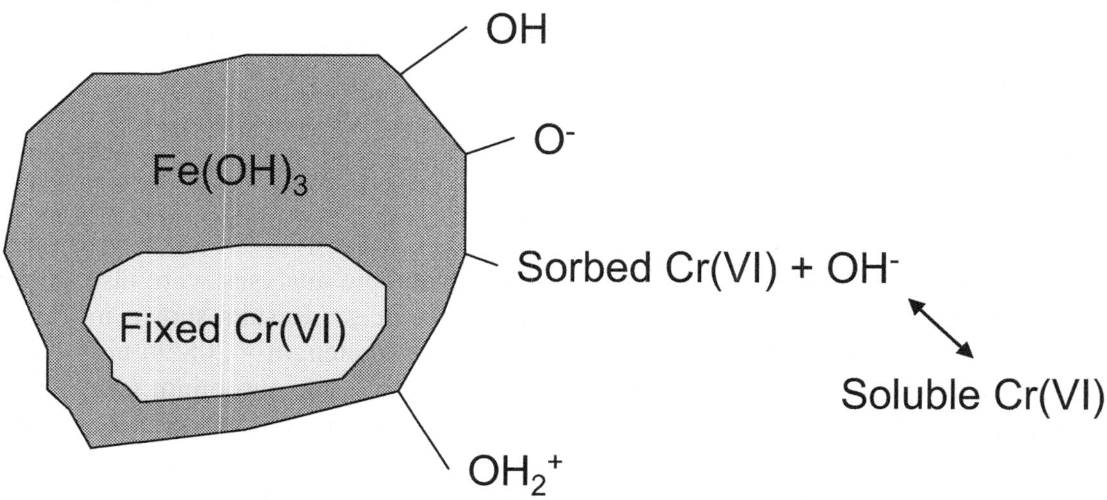

Figure 2-1. Three fractions associated with iron hydroxide: soluble, sorbed, and fixed

Methods

Synthetic waters were prepared to evaluate boron and chromium recovery in the presence of iron particles as described in Table 2-5.

Table 2-5
Synthetic waters used to evaluate boron and chromium recover in the presence of iron particles

Water	Fe Condition	Fe* mg/L	pH	B µg/L	Cr(VI) µg/L	Cr(III) µg/L
1	Preformed	10	5.0	200		
2	In-situ	10	5.0	200		
3	Preformed	5	5.0		40	
4	In-situ	5	5.0		40	
5	Preformed	5	8.4		40	
6	In-situ	5	8.4		40	
7	Preformed	5	5.0			40
8	In-situ	5	5.0			40
9	Preformed	5	8.4			40
10	In-situ	5	8.4			40

* Iron was dosed as ferric to achieve the described concentrations as Fe.

In these experiments, the synthetic waters initially adjusted to the prescribed pH conditions. Two iron conditions were used in the experiments:

- For the "preformed" iron condition, ferric iron was dosed to the water along with a predetermined amount of sodium hydroxide to maintain the desired pH. After 30 minutes, chromium or boron was spiked into the solution. Iron particles formed during this experiment were termed "preformed," since the ferric hydroxide particulate was formed before chromium or boron was added.
- For the "in-situ" iron condition, ferric iron was dosed after chromium or boron was spiked allowing iron particles to form "in-situ" with chromium or boron already present in the water. It was expected that chromium in particular might be more difficult to recover from "in-situ" iron particles since it could be incorporated into the solid structure as fixed Cr or a Fe(III)-Cr(III) hydroxide precipitate (Sass and Rai, 1987).

Solutions were allowed to react for approximately one hour. Aliquots from the waters containing boron were handled in three ways: (1) alkaline digested with 12 mM soda ash, (2) alkaline digested with 12 mM soda ash and 0.45 µg/L filtered, and (3) acid digested with 5% nitric acid. These samples were analyzed for total boron by ICP-ES following Standard Method 3120 B (APHA, AWWA, WEF 1998). Aliquots from the waters containing chromium underwent more rigorous collection, digestion, and analysis according to the procedures described in Table 2-6.

Table 2-6
Chromium sample generation and processing procedures for iron solids experiment

Procedure	Descriptor	Filtered (0.45 μm)	Initial Treatment	Digestion	Sample Analysis	Comments
1	S.M. 3030-A (dissolved metal)	Yes	Acidify to pH 2	Sit stagnant for 24 h	Cr(VI) by IC Total Cr by ICP-MS	Valid for dissolved Cr if turbidity <1 NTU
2	S.M. 3030 A (total metal)	No	Acidify to pH 2	Sit stagnant for 24 h	Cr(VI) by IC Total Cr by ICP-MS	Valid for total Cr if turbidity <1 NTU
3	USEPA Method 1639-12.2.7 (USEPA 1996b)	No	Add 0.3 mL HNO_3 and 0.15 mL HCl to 30 mL aliquot	Heat to 85°C for 2 h; 0.45 μm filter; acidify to 2% HNO_3	Total Cr by ICP-MS	Sample digestion was not performed in bottle as called for in method
4	S.M. 3030-E	No	Add 0.3 mL HNO_3 and 0.15 mL HCl to 30 mL aliquot	Heat to 105°C in block heater for 2 h; 0.45 μm filter; acidity to 2% acid by adding 0.15 mL HNO_3	Total Cr by ICP-MS	Sample filtered instead of centrifuged as called for in method
5	Peroxide digestion	No	Add 0.3 mL H_2O_2 to 30 mL aliquot	Sit stagnant for 24 h; 0.45 μm filter, acidify to 2% HNO_3	Total Cr by ICP-MS	Peroxide may aid in the recovery of total Cr
6	S.M. 3500-Fe B	No	Add 2 mL HCl and 1 mL hydroxylamine solution (10 g NH2OH·HCl diluted to 100 mL) to 50 mL aliquot	Heat to 85°C for 2 h; 0.45 μm filter	Total Cr by ICP-MS	Sample not allowed to boil per method; iron not analyzed with phenanthroline

Results

In all of the samples analyzed for boron, all of the initially-spiked boron was recovered and therefore determined to remain soluble; i.e. no boron sorbed to or was incorporated into any iron particulate in either the "preformed" or "in situ" solutions.

For the chromium waters, the solutions containing "in situ" and "preformed" iron hydroxide solids synthesized at pH 5 had initial turbidities less than 1 NTU (0.35 and 0.80, respectively), while both waters at pH 8.4 had turbidities of 1.4 NTU.

Standard Method 3500-Fe B recommends a hydroxylamine digestion to measure total iron in an environmental sample. Therefore, it was anticipated that the hydroxylamine digestion procedure would give the highest recoveries for Cr since it dissolves the iron particulates, thereby releasing any "fixed" chromium. Recoveries of total chromium per the hydroxylamine digestion procedure were indeed the highest and ranged between 90% and 104%. Iron recoveries in these samples were also high at 89% to 97%. In comparison, without the addition of hydroxylamine, recoveries of chromium were as low as 21% for the peroxide digestion and as low as 83% for hot HNO_3 digestion (Standard Method 3030E). Chromium recoveries by Standard Method 3030E were significantly lower ($P < 0.0032$) than for hydroxylamine digestion.

Per the standard EPA protocol for total chromium determination, the waters with turbidity less than 1 NTU could have been acidified to pH < 2 and analyzed directly. Following this procedure the recovery of chromium was 85% for the "preformed" water (by comparison, using Standard Method 3030E chromium recovery was also 85%). For the "in situ" water, recovery of total Cr could not be accurately determined on the ICP-MS due to an interference with the ICP-MS analysis for chromium. Interference from particulate material in the sample is the most likely cause as described earlier. Iron recovery from the pH 5 "preformed" water was only 92% using Standard Method 3030E.

Strictly following EPA protocol, a Standard Method 3030E nitric acid digestion was required for the pH 8.4 samples (turbidities = 1.4 NTU). The recovery of total chromium was 83% - 89% while iron recoveries were 76% ("preformed") and 91% ("in situ").

These experiments show that following the EPA protocol could lead to at least a slight underestimation of the amount of total chromium in a water sample with fresh iron hydroxide. A fairly large amount of particulate material can be present in a water sample even if turbidity is less than 1 NTU. If this material contains iron, chromium may be adsorbed to the particles or it may be fixed such that it will not be released into solution at pH 2. In addition, Standard Method 3030 E cannot always dissolve even relatively fresh iron hydroxide in samples with turbidity above or below 1 NTU. In some instances even filtration and acidification cannot remove all particulate matter that may cause problems. In such cases hydroxylamine digestion led to improved recovery of both chromium and iron. These experiments led to the final sampling protocol for the survey which utilized hydroxylamine digestion for improved chromium recovery.

EVALUATION OF SAMPLE PRESERVATION IN SIMULATED NATURAL WATERS

Six waters were prepared to simulate a range of natural water conditions that might be encountered in the survey. Two liters of each water were prepared as described below and periodic sampling and analysis was performed to determine preservation effects and the viability of the sample processing and analytical protocol. The purpose of this test was two-fold: (1) to determine the effectiveness of sample preservation for maintaining chromium speciation, and (2) to determine if boron and chromium can be accurately measured in a variety of raw water sample conditions.

Water Preparation

The six (6) waters were prepared as follows:

1. **Salt only:** This water contained 1 mM NaCl and 0.5 mM $NaHCO_3$.
2. **Particulate:** This water was identical to (1) with bentonite, natural organic matter (NOM), and ferric iron added. Bentonite was added to give a turbidity of about 100 NTU. NOM was added in the form of 3 mg/L fulvic acid. Also, 3 mg/L ferric iron was added and expected to form ferric hydroxide.
3. **Reducing:** This water contained 1 mM NaCl, 0.5 mM $NaHCO_3$, 2 mg/L ferrous iron, 0.3 mg/L manganese, and 1 mg/L hydrogen sulfide.
4. **Hard groundwater:** This water contained 1 mM NaCl, 0.5 mM $NaHCO_3$, 300 mg/L (as $CaCO_3$) calcium and 100 mg/L (as $CaCO_3$) magnesium. Additional alkalinity in the form of 3 mM sodium carbonate was also added. Carbon dioxide was bubbled through this water to lower the pH to 6 before calcium and magnesium were added to prevent the formation of calcite.
5. **Chlorine:** This water contained 1 mM NaCl, 0.5 mM $NaHCO_3$, and 5 mg/L free chlorine. A small amount of bentonite was also added to promote any surface catalyzed reactions that might occur between chlorine and chromium.
6. **Biologically active:** This water was identical to (1) but with 20 mL of activated sludge from sewage treatment also added.

An identical water to each of the above was also prepared containing 12 mM soda ash (Na_2CO_3). In each of these waters, the preservative was added to the water prior to the addition of boron and chromium. This was the preservative dose used for the final occurrence survey. Table 2-7 displays the initial pH conditions for the unpreserved and preserved waters.

Table 2-7
Initial pH conditions of simulated natural waters

Water	Unpreserved Initial pH	Preserved Initial pH
Salt only	7.95	10.97
Particulate	3.75	10.78
Reducing	7.28	10.95
Hard groundwater	6.46	9.89
Chlorine	8.49	10.96
Biologically active	8.40	10.89

After preservation, the pH conditions of the water samples ranged were all well within the range of the alkaline preservation target.

Experimental Procedure

Each water was sampled immediately before and after spiking 100 µg/L boron, 40 µg/L Cr(III), and 40 µg/L Cr(VI). Each water was also sampled after sitting stagnant for one week and two weeks.

The preserved waters (12 mM soda ash) were prepared so that the effectiveness of the alkaline digestion and the preservative could be evaluated. The preserved waters were processed in two ways. The first aliquot was filtered and analyzed without digestion. The second aliquot was not filtered and then alkaline digested per a modified version of EPA Method 3060A – alkaline digestion for hexavalent chromium (USEPA 2000). The $NaOH/Na_2CO_3$ digestion reagent was added in a 20 mL per 100 mL sample ratio. The resulting solution was placed in a sonicator/hot water bath at 60°C for 90 minutes instead of 95°C for 60 minutes as described by the method. Each aliquot was then filtered and analyzed for boron, total chromium, and Cr(VI).

The waters labeled as unpreserved (without 12 mM soda ash) were sampled in accordance with the processing and analytical protocol initially designed to be used in the survey. Samples from the unpreserved waters were preserved by the addition of 12 mM soda ash (Na_2CO_3) immediately after each sample was withdrawn. These samples were processed in three ways. The first aliquot was filtered and alkaline digested as described above. The second aliquot was not filtered, but was alkaline digested. The third aliquot not filtered, but was acid digested by adding 1 mL concentrated nitric acid per 50 mL sample. The third aliquot was also placed into a sonicator/hot water bath at 60°C for 90 minutes. Each aliquot was then filtered and analyzed for boron, total chromium, and Cr(VI) with the exception of the acid digested samples which were not analyzed for Cr(VI).

Results

The complete sample analysis results for boron, total chromium, and hexavalent chromium from these experiments are contained in Appendix A of this report.

Total boron analyses were unaffected by preservation and digestion procedures evaluated in these experiments. Total boron recovery ranged from 97.6% to 108.8% at two weeks in preserved, unfiltered, acid-digested aliquots of all waters tested. In the preserved, unfiltered, alkaline-digested aliquots, total boron recovery ranged from 87.3% to 102.7% at two weeks. Soluble boron recovery – determined from the filtered, alkaline-digested aliquot –ranged from 95.6% to 100.7%. These data indicate the vast majority of boron remained soluble throughout the experiments and boron was not sensitive to the preservation technique. The implication of this result was that preservation would not affect boron concentrations determined in the final occurrence survey samples.

Total chromium recovery by acid digestion for unpreserved samples was effective for most water conditions over the two week testing period; however, the total chromium concentration decreased significantly in the "particulate" water. Total chromium concentrations also decreased in the "biologically active" water, but to a lesser extent. A possible explanation is that the chromium is slowly incorporated into particulate matter and the acid digestion employed in this phase of work does not recover it all.

Preservation of soluble Cr(VI) using 12 mM soda ash appears to be sufficient for the entire range of waters tested. In preserved, filtered, undigested aliquots, 39.7 µg/L to 49.7 µg/L of Cr(VI) was measured at two weeks. Considering the original 40 µg/L Cr(VI) spike, these results may imply that some of the Cr(III) spike was oxidized to Cr(VI) after preservation. However, this is more likely an artifact of the experimental protocol.

As was the case in the previous material tests, total Cr(III) is not recovered from the high pH solutions due to apparent sorption to the bottles. However, total recoverable Cr(III) can still be estimated as the difference in total chromium and Cr(VI) analysis of digested samples.

There were several interesting observations concerning the chromium speciation data. For the "salt only", "hard groundwater", and "biologically active" waters, over 80% of the 40 µg/L Cr(VI) spike was recovered from the preserved, filtered aliquots and deemed soluble. No chromium – either total or Cr(VI) – was recovered from the "particulate" water after the alkaline digestions, demonstrating that chromium concentrations cannot be accurately determined for samples with similar characteristics using these digestion techniques. These results imply that, regardless of preservation, 100% of the chromium will not be recovered from a water sample that has appreciable particulate matter; especially if the chromium has coprecipitated with that particulate. This gave rise to consideration of more rigorous sample digestion procedures.

Contrary to expectations, the majority of the Cr(VI) spike was recovered even after two weeks in the presence of reducing conditions even without preservation, indicating that no reduction of Cr(VI) to Cr(III) took place. Further calculations revealed that the solubility product of FeS would be exceeded at pH 7 for the concentrations of ferrous iron and sulfide initially spiked and observations during the sampling confirmed a yellow-brown particulate during sample filtration. This may also be explained as ferrous oxidation by dissolved oxygen. Therefore, precipitation of the reducing agents explains the lack of chromium reduction.

The "chlorine" water appeared to oxidize much of the Cr(III) to Cr(VI) if the sample was unpreserved, but very little oxidation occurred in alkaline preserved samples. These results imply that preservation stops redox changes of chromium due to chlorine.

EVALUATION OF FLUORIDE COMPLEXATION WITH BORON

Boron can react with fluoride to form a volatile complex, BF_3, which has a boiling point of -101 °C (Weast, Astle, and Beyer 1988). This could potentially lead to the loss of boron in the occurrence survey samples. Wood and Nicholson performed a test at fairly high levels of boron (0.1, 1.0, and 10.0 mg/L) and fluoride (0.7, 7.0, and 70.0 mg/L) and found that there was a 16% decrease in boron concentration over a 14 day period (Wood and Nicholson, 1995). During the QA/QC portion of this study an experiment was constructed to test whether this volatile complex could form in a reasonable length of time in waters containing less than the EPA's maximum contaminant level (MCL) of fluoride of 4 mg/L.

Method

Four waters were prepared with 1 mM sodium bicarbonate and 200 µg/L boron. Two of the waters were adjusted to pH 7 with HCl and the other two to pH 10.7 by adding 12 mM soda ash. Fluoride (4 mg/L, 11.3:1 F:B molar ratio) was added to one of the pH 7 waters and one of the pH 10.7 waters. The waters were allowed to sit overnight to give the B-F complex an opportunity to form. Air was then bubbled through each of the waters for three hours.

Results

The final boron concentrations were 5% to 10% higher than the amount originally spiked into each of the four waters. There was no appreciable difference in boron concentrations between samples with and without fluoride. Loss of boron is not anticipated due to fluoride complexation

CONCLUSIONS FROM THE ANALYTICAL METHODS INVESTIGATION

Boron remains soluble in all types of waters: waters with substantial particulate matter, waters with sulfides present, waters with a large amount of alkalinity, and waters with chlorine present. Additionally, all of the boron present in waters under these extreme conditions can be recovered using the final sampling protocol (see below).

Two types of error have been identified in determination of total chromium using standard techniques – both arising from interactions with iron particles that are present in many ambient waters and recalcitrant to dissolution in weak (pH 2) or strong (5%) acid digestion.

1. If a sample collected in the field is acidified to slightly less than pH 2.0, soluble Cr(VI) can sorb to iron particles if they are present. These particles can be lost from solution by a variety of mechanisms (attachment to the sides of the container or sedimentation in test tubes), and the Cr(VI) recovery can approach 0%.
2. Chromium in samples can be present in a "fixed" form. This refers to Cr associated with particles that is not solubilized during a nitric acid digestion. This type of Cr will not be recovered by Standard Method 3030E; however, a much greater portion of this chromium is recovered by hydroxylamine digestion.

FINAL SURVEY SAMPLE PRESERVATION, DIGESTION, AND ANALYSIS METHODS

The final survey sample preservation, digestion, and sample analysis procedures were developed based on results from this investigation of both chromium and boron analytical challenges. The final survey sample kit included two bottles: one (Bottle A) acid-preserved and the other (Bottle B) base-preserved. Sample kits were sent from Virginia Tech to participating drinking water utilities that collected samples following detailed sampling instructions. After samples kits were returned to Virginia Tech by the participating utilities, the acid-preserved sample was digested in the bottle with hydroxylamine following a procedure similar to that specified in Standard Method 3500 – Fe B (APHA, AWWA, WEF 1998). Following this procedure, 4% hydroxylamine and 2% hydrochloric acid were added to the bottle. The bottle was then heated to 85°C for 24 hours. The resulting digested solution was then filtered using a 0.45 μm pore size filter and analyzed for total boron by ICP-ES following Standard Method 3120 B (APHA, AWWA, WEF 1998) using a JY Ultima ICP-ES at Virginia Tech. An aliquot of the hydroxylamine digested and filtered sample was sent to Utah State University (USU) for total chromium analysis by ICP-MS following EPA Method 200.8 (Determination of Trace Elements in Waters and Wastes by Inductively Coupled Plasma – Mass Spectrometry) (USEPA 1994) using an Agilent 7500c ICP-MS with an octopole reaction system. The base-preserved sample was filtered in the Virginia Tech lab and then sent to USU for hexavalent chromium analysis by IC following EPA Method 1636 (Determination of Hexavalent Chromium by Ion Chromatography) (USEPA 1996a) on a Dionex DX-320 IC system with an AD25 UV-Vis detector. Figure 2-2 describes the sample kit preservation and subsequent digestion, filtration, and sample analysis procedures used for the regular occurrence survey samples.

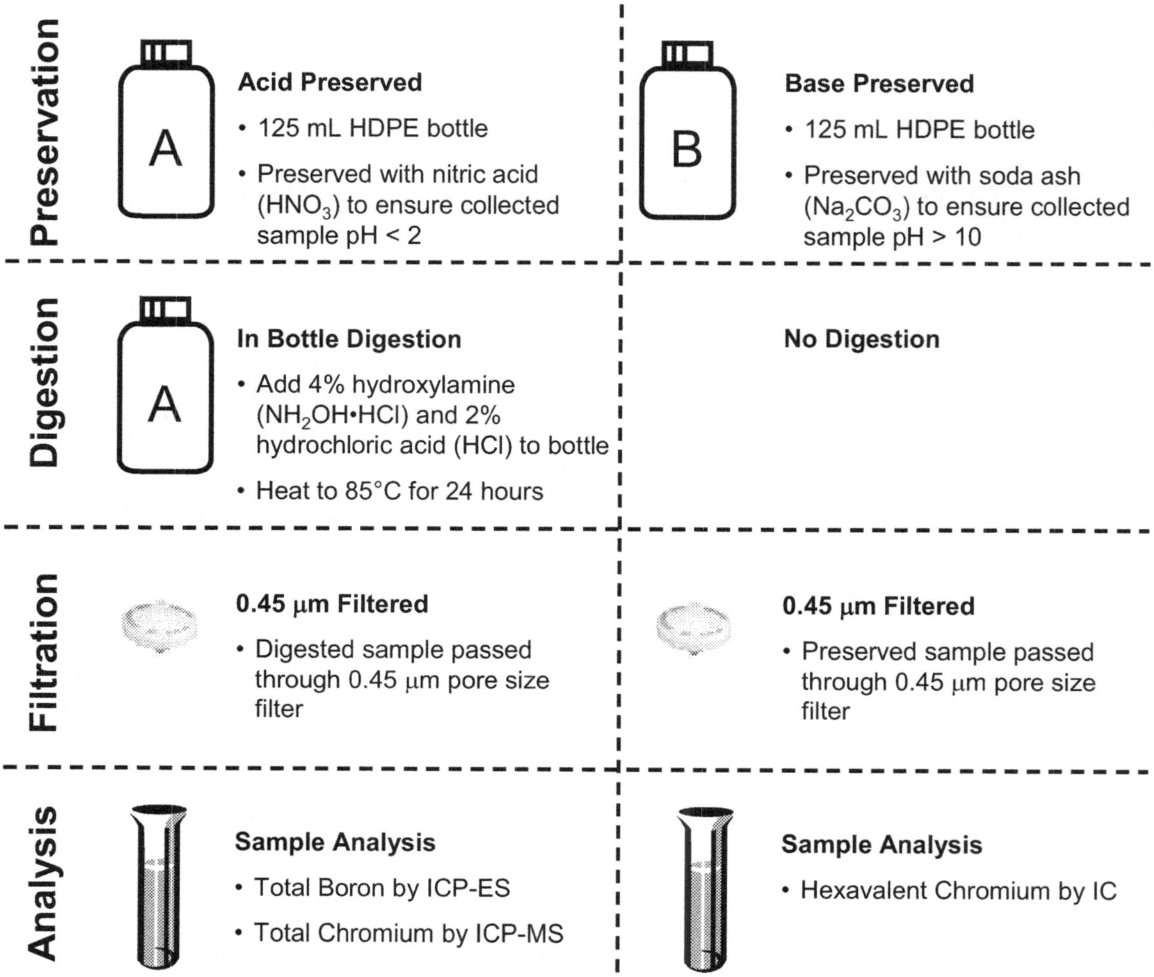

Figure 2-2. Sample collection and subsequent digestion and analysis used for the regular occurrence survey samples

CHAPTER 3
REVIEW OF EXISTING OCCURRENCE DATA SOURCES

IDENTIFICATION AND EVALUATION OF EXISTING OCCURRENCE DATA SOURCES

Boron and chromium occurrence information was gathered from available databases and previous surveys to guide the development and design of the national occurrence survey. The intent of this *a priori* assessment was to identify general occurrence trends of boron and chromium and perhaps determine the sample size required to capture regional occurrence patterns.

Several data sources were identified and evaluated for historical chromium and boron occurrence data in support of this project. These data sources are managed by the United States Geological Survey (USGS), the United States Environmental Protection Agency (USEPA), and the State of California Department of Health Services (CalDHS). USGS manages and maintains many water quality related databases. They include: historical NWIS-I and WATSTORE, the current up-to-date NWIS-II (NWISWeb), the National Water Quality Assessment (NAWQA), and the National Stream Water-Quality Monitoring Networks (WQN) which include the National Stream Quality Accounting Network (NASQAN) and Hydrologic Benchmark Network (HBN). USEPA manages water quality information in the STORET data system. The USEPA also has other occurrence information of relevance to this project including the National Inorganic and Radionuclide Survey (NIRS) and the 16-State occurrence data (16-State) in support of the Six-year Review of the Existing National Primary Drinking Water Regulations. Finally, the State of California Department of Health Services (CalDHS) manages and maintains drinking water quality data in their own Water Quality Monitoring Database.

Each of the identified occurrence data sources were evaluated to determine their suitability for providing meaningful occurrence information about boron, total chromium and hexavalent chromium. This evaluation took into consideration the following factors:

- **Data availability** – Were the data readily available in an electronic format accessible by the project team?
- **Geographic coverage** – How were the sample results geographically distributed?
- **Period of record** – When were the sample results generated?
- **Source waters types sampled** – Did the sample results correspond to either surface water (SW) or groundwater (GW)? Were the samples collected from drinking water sources or ambient waters not necessarily used as drinking water sources?
- **Boron data** – Did the data source include boron sample results?
- **Total chromium data** – Did the data source include total chromium sample results?
- **Hexavalent chromium data** – Did the data source include hexavalent sample results?

Based on these factors, each data source was determined to be:

1. infeasible for use
2. of limited relevance to the project, or
3. of meaningful relevance to the project.

For instance, if a data source was not readily available it was considered infeasible for use. Data sources were considered of limited relevance if they were readily available, but were limited based on some other factor such as geographic coverage, source water types sampled, or specific occurrence data. Table 3-1 summarizes the results of the data source evaluations.

Table 3-1
Evaluation of existing occurrence data sources

Data Source	USGS	USGS	USGS	USGS	USEPA	USEPA	USEPA	CalDHS
Database	**NWIS-I**	**NWIS-II**	**NAWQA**	**WQN**	**STORET**	**NIRS**	**16-State**	**Water Quality Monitoring Database**
Data Availability	Available on EarthINFO CD-ROMs.	Limited querying and downloading online. National data downloads not possible at the time of the evaluation.	Available online.	Available online.	Available online. National data downloads are not practical because of limited query tools.	Obtained directly from USEPA.	Obtained directly from USEPA.	Obtained directly from California DHS.
Geographic Coverage	50 States and Puerto Rico	50 States and Puerto Rico	50 Study Units to be completed.	HBN: 63 small watersheds. NASQAN: 618 larger watersheds.	Random coverage of all 50 States, territories, and jurisdictions of the U.S.	49 States (Hawaii excluded)	16 States: AL, CA, FL, IL, IN, KY, MI, MT, NE, NJ, NM, OR, SC, SD, TX, VT	California. drinking water compliance data
Period of Record	Early 1900s - 1998	Historical WATSTORE data is currently being added. Current and future data added periodically.	Data collection is dependent on study unit: 1991, 1994, 1997, or 1999.	HBN: 1962-95 NASQAN: 1973-95	1900s - Current	1984 - 1986	1987 - 1998	1980s – 2003
Source Water Sampled	Ambient SW and GW	Ambient SW and GW	Ambient SW and GW	Ambient SW	Ambient SW and GW	Drinking Water GW	Drinking Water SW and GW	Drinking Water SW and GW
Boron Data?	Yes	Yes	Yes	Yes	Not determined	Yes	No	Yes
Total Chromium Data?	Yes	Yes	Yes	Yes	Not determined	Yes	Yes	Yes
Hexavalent Chromium Data?	Yes	Yes	No	No	Not determined	No	No	Yes
Data Relevance	Meaningful	Infeasible	Limited	Limited	Infeasible	Meaningful	Limited	Meaningful

As seen in Table 3-1, the USGS NWIS-I, USEPA NIRS, and CalDHS Water Quality Monitoring databases were identified as meaningful data sources for this project. These three data sources were further analyzed to provide baseline occurrence information for the project.

ANALYSIS OF MEANINGFUL OCCURRENCE DATA SOURCES

Each of the three "meaningful" data sources is unique and provides a different perspective of occurrence information. The following sections describe and compare each of these data sources in more detail. In particular, detection limits for each parameter are discussed with respect to each data source. Both the USGS NWIS-I and CalDHS data sets contain sample results collected over a long timeframe. These data sets include multiple detection limits for individual parameters – as expected with methods improving over the course of the sampling program. In an effort to equally present data and provide context for differences in low-level sample results from each source, data set detection limits are described for each parameter for each data set. When presenting occurrence results from each data set in this report, below detection limit results are presented as zeros.

USGS NWIS-I

The USGS NWIS-I database contains the greatest number of records over the longest timeframe for boron, total chromium, and hexavalent chromium. However, these results represent ambient source water samples that may not necessarily reflect drinking water sources. This database is limited in that its data coverage is not equally distributed geographically or temporally. The numbers of sample results also vary by sampling location. For instance, many sampling locations with non-detectable concentrations of boron or chromium were sampled just once, while locations with elevated concentrations were repeatedly sampled. In addition to this, many sample results contained in the data set are the result of targeted sampling programs in specific states or regions for specific parameters. Both of these practices can bias the overall occurrence trends for the data set.

Analysis of data from this source was challenging because of detection limit issues. It contains sample results collected over a long timeframe resulting in a number of detection limits applied throughout the data set. In order to understand the historical detection limits used to generate the sample results, data set detection limits were established for each parameter. To do so, records containing below detection limit (BDL) flags were counted by their associated detection limits. Then cumulative probabilities were assigned to each of the count values for each detection limit. Data set detection limits were assigned as the lowest BDL value that represents more than 50% of all BDL counts. Defining data set detection limits allows this data source to be evaluated in context and equally with other occurrence data sources. Figure 3-1 displays an example of the resultant data for groundwater total chromium results.

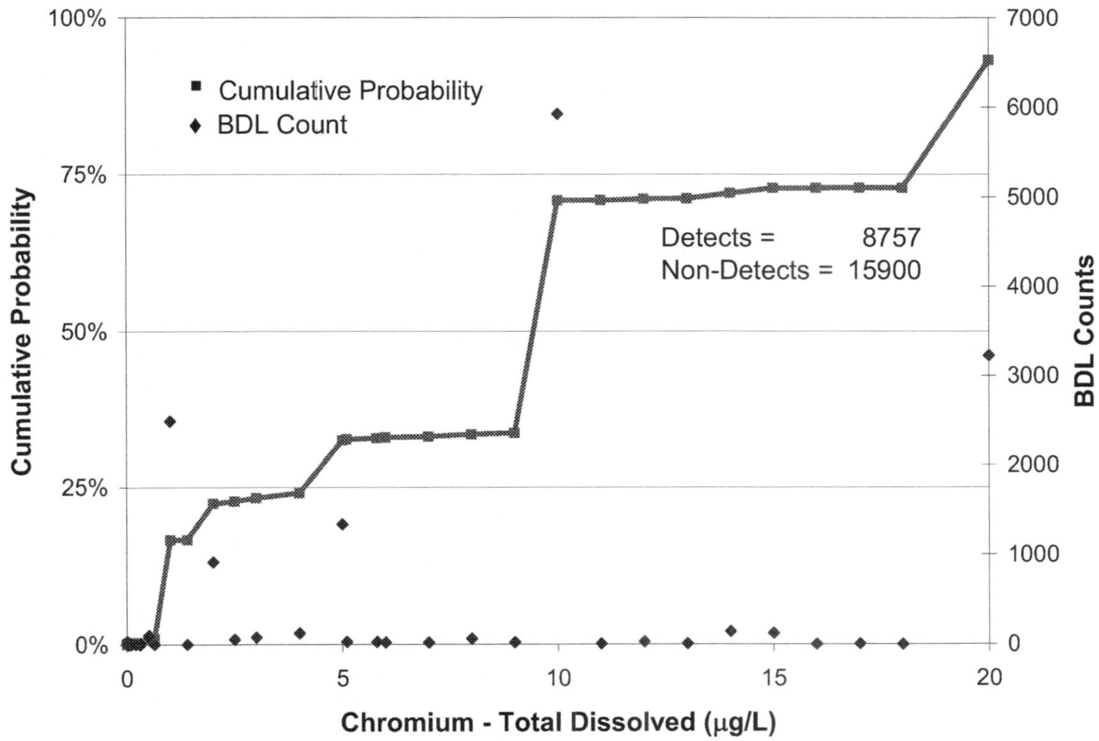

Figure 3-1. Detection limit assessment for USGS NWIS-I groundwater total chromium results

As seen in Figure 3-1, the lowest BDL value that represents more than 50% of all BDL counts is 10 µg/L. Therefore, the data set detection limit for groundwater total chromium results was set to 10 µg/L. Table 3-2 displays the data set detection limits for boron, total chromium, and Cr(VI) for both surface water and groundwater sources for the USGS NWIS-I data source. Counts of both BDL values and detected values are also shown.

Table 3-2
Data set detection limits by parameter and source water for USGS NWIS-I data

Parameter	Source Water	Data Set Detection Limit (µg/L)	BDL Count	Detected Count
Boron	SW	50	12,311	167,662
	GW	20	4,466	76,364
Total Chromium	SW	5	64,305	30,668
	GW	10	15,900	8,757
Cr(VI)	SW	1	3,755	11,371
	GW	1	1,292	11,337

USEPA NIRS

The National Inorganic and Radionuclide Survey (NIRS) was designed and conducted by EPA in the mid-1980s specifically to provide data on the occurrence of a select set of radionuclides and inorganic chemicals being considered for National Primary Drinking Water

Regulations. It contains boron and total chromium data, but not hexavalent chromium data. A detailed description of the survey can be found in Longtin (1988). NIRS was structured as a stratified, random sampling of the nation's community groundwater supplies as they existed in the early 1980s. The survey involved the selection of 1,000 ground water systems. At the time that the NIRS samples were selected, there were approximately 48,000 community groundwater systems. The 1,000 systems were selected randomly from within four size-based stratification levels: 25 to 500; 501 to 3,300; 3,301 to 10,000; and greater than 10,000. Including 1,000 systems in the survey provided for selection of approximately 2% of the systems within each of the four size categories.

As a controlled occurrence survey, the NIRS data have consistent detection limits for each parameter. Table 3-3 displays the data set detection limits for boron and total chromium for groundwater sources for the USEPA NIRS data source. Counts of both BDL values and detected values are also shown.

Table 3-3
Data set detection limits by parameter and source water for EPA NIRS data

Parameter	Source Water	Data set Detection Limit (μg/L)	BDL Result Count	Detected Result Count
Boron	GW	5	179	810
Total Chromium	GW	2	926	63

CalDHS Water Quality Monitoring Database

The State of California Department of Health Services (CalDHS) manages and maintains drinking water quality data in their own Water Quality Monitoring Database. This data source contains drinking water compliance data for boron, total chromium, and hexavalent chromium with significant geographical coverage of California. Both raw and treated water occurrence information is available by location depending on whether treatment is installed for the given contaminant's removal. This data source includes data beginning in the 1980s and is updated regularly.

While this data source is limited geographically by only covering California, California has been very proactive in sampling for hexavalent chromium. A hexavalent chromium sampling program was initiated by the state in 2000. This sampling program provides detailed hexavalent chromium occurrence data for California drinking water sources.

The CalDHS data source is similar to the USGS NWIS-I; numerous detection limits are applied throughout the dataset for each parameter. The same data set detection limit evaluation use for the NWIS-I data was also completed for the CalDHS data. Table 3-4 displays the data set detection limits for boron, total chromium, and Cr(VI) for both surface water and groundwater sources for the CalDHS data source. Counts of both BDL values and detected values are also shown.

Table 3-4
Data set detection limits by parameter and source water for CalDHS data

Parameter	Source Water	Data set Detection Limit (µg/L)	BDL Count	Detected Count
Boron	SW	100	166	1,026
	GW	100	1,729	5,403
Total Chromium	SW	10	2,187	1,500
	GW	10	15,063	9,980
Cr(VI)	SW	1	33	169
	GW	1	213	2,220

While the data set detection limits for the CalDHS data are 100, 10, and 1 µg/L for boron, total chromium, and Cr(VI), the data set contains many measurable (i.e. non-BDL) sample results below those limits.

SUMMARY OF MEANINGFUL OCCURRENCE DATA SOURCES

It is important to understand the boron, total chromium, and hexavalent chromium occurrence data from each meaningful data source. Figures 3-2, 3-3, and 3-4 display cumulative probability distributions for each parameter by source water type from each meaningful data source available. Below detection limit (BDL) results are presented as zeros.

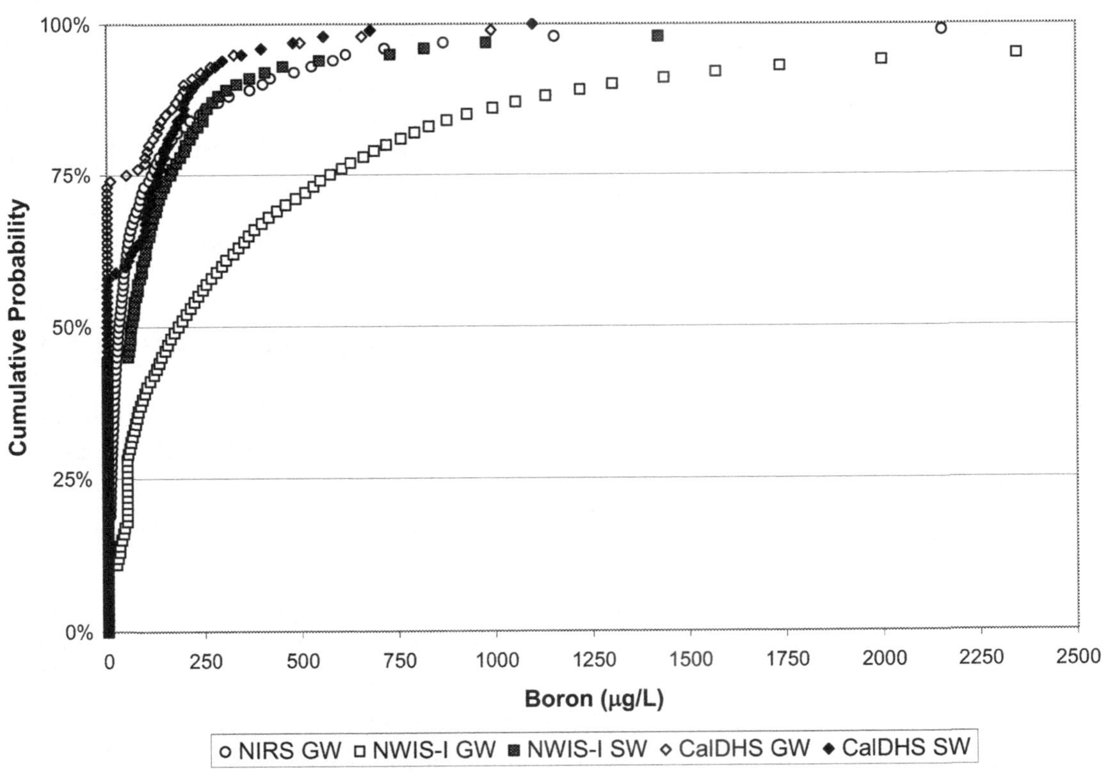

Figure 3-2. Cumulative probability distributions of boron occurrence data

There are clear differences between the boron occurrence data found in the NIRS or CalDHS and NWIS-I data sets. NIRS, which sampled 1,000 groundwater drinking water sources, shows significantly lower boron occurrence than found in the ambient groundwater samples shown in the NWIS-I data set. In fact, the NIRS groundwater data more closely resemble the NWIS-I surface water data. The CalDHS data shows even lower boron occurrence than the NIRS. These trends are related to the sampling bias of the CalDHS and NIRS data sets where only drinking water sources were sampled. It would be expected that the NIRS and CalDHS drinking water sources would have lower boron concentrations since waters (particularly groundwaters) high in total dissolved solids (TDS) and associated high boron concentrations would not generally be utilized as drinking water sources.

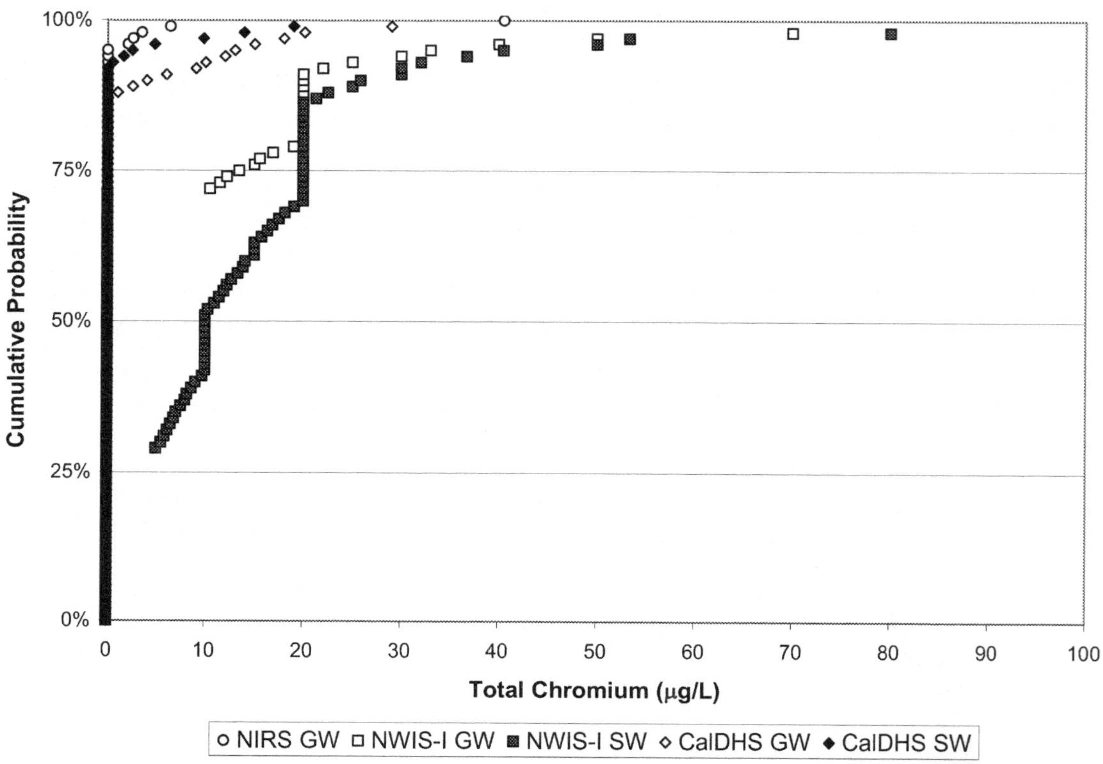

Figure 3-3. Cumulative probability distributions of total chromium occurrence data

The total chromium occurrence data presented in Figure 3-3 show that groundwater and surface water total chromium occurrence data from NWIS-I are significantly higher than the drinking water source occurrence data from NIRS and CalDHS. When comparing NIRS and CalDHS, the California occurrence data – both surface and groundwater – is higher than the national groundwater data.

The high detection limits associated with the NWIS-I data can be seen in Figure 3-3 where the cumulative probabilities follow the 0 μg/L line until reaching 5 μg/L for surface water and 10 μg/L for groundwater. The NWIS-I dataset also contains a significant number of detected sample results at 10 μg/L and 20 μg/L that were not flagged in the NWIS-I database as BDL results. These results are related to the historical nature of the NWIS-I data set and show that the ambient water quality results from NWIS-I may not be representative of drinking water source water quality.

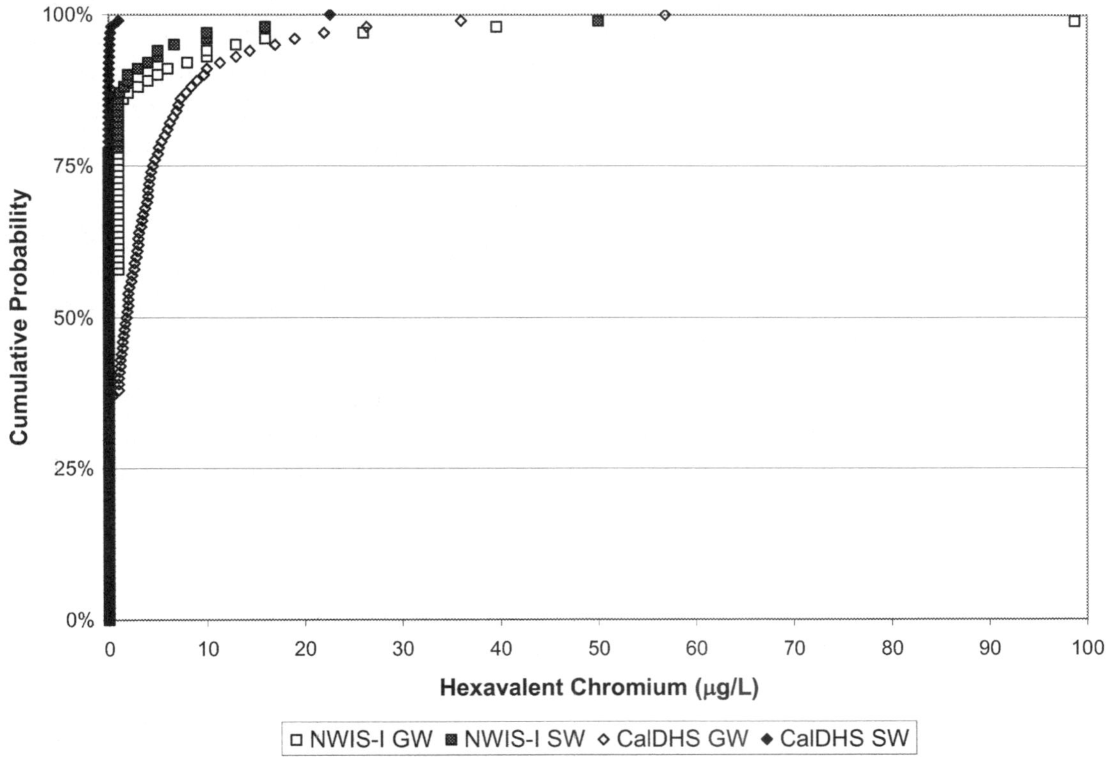

Figure 3-4. Cumulative distributions of hexavalent chromium occurrence data

Figure 3-4 shows that the CalDHS hexavalent chromium data is significantly different for surface and groundwaters. The NWIS-I data, on the other hand, indicates no real difference between surface and groundwater sources. These hexavalent chromium data can be also be compared to the total chromium data in Figure 3-3. On this comparative basis, the NWIS-I hexavalent chromium data – for both groundwater and surface water – are only slightly higher than the NIRS groundwater total chromium data. The NWIS-I hexavalent chromium data are significantly lower than the corresponding NWIS-I total chromium data. The CalDHS groundwater hexavalent chromium data are significantly higher than the associated total chromium results.

CONCLUSIONS

Eight data sets were identified and evaluated to determine their suitability for providing meaningful occurrence information about boron, total chromium, and hexavalent chromium. Of these eight data sets, the USGS NWIS-I, USEPA NIRS, and CalDHS Water Quality Monitoring databases were identified as meaningful data sources for this project and warranted further analysis.

Each of these three data sets provides different perspectives of boron, total chromium, and hexavalent chromium occurrence. NWIS-I contains the most sample results for each parameter (more than 10,000 records per parameter). However, these results represent ambient water sources that are not necessarily reflective of drinking water sources. The NWIS-I data set also has challenging detection limit issues.

The NIRS was designed and conducted by EPA in the mid-1980s specifically to provide data on the occurrence of a select set of radionuclides and inorganic chemicals being considered for National Primary Drinking Water Regulations. It contains boron and total chromium data, but not hexavalent chromium data. NIRS was structured as a stratified, random, statistical survey of approximately 48,000 groundwater drinking water sources resulting in samples of 1,000 groundwater systems.

The CalDHS Water Quality Monitoring Database provides boron, total chromium, and hexavalent chromium data for drinking water sources in California. It is unique in that it contains data from the hexavalent chromium sampling program initiated in 2000. These results are directly comparable with the results from this survey project.

The occurrence results from the meaningful occurrence data sources indicate differences between ambient source water quality and drinking source water quality. Therefore, the NWIS-I data may not be suitable for anticipating drinking source water quality in the absence of such data.

CHAPTER 4
DESIGN OF NATIONAL OCCURRENCE SURVEY

RECRUITMENT OF PARTICIPATING UTILITIES

Drinking water utilities were recruited to participate in the NCBOS throughout the duration of the project. To start, 20 utilities committed to participating in the project and give in-kind support at the time of the proposal submittal. Active recruitment of utilities during the project occurred in two phases: (1) indirect contact through online and printed screening surveys, and (2) direct contact by telephone.

The first recruitment phase solicited utility participation in the survey using both online and print screening surveys. Figure 4-1 displays the front page of the online screening survey; and Figure 4-2 shows the print screening survey.

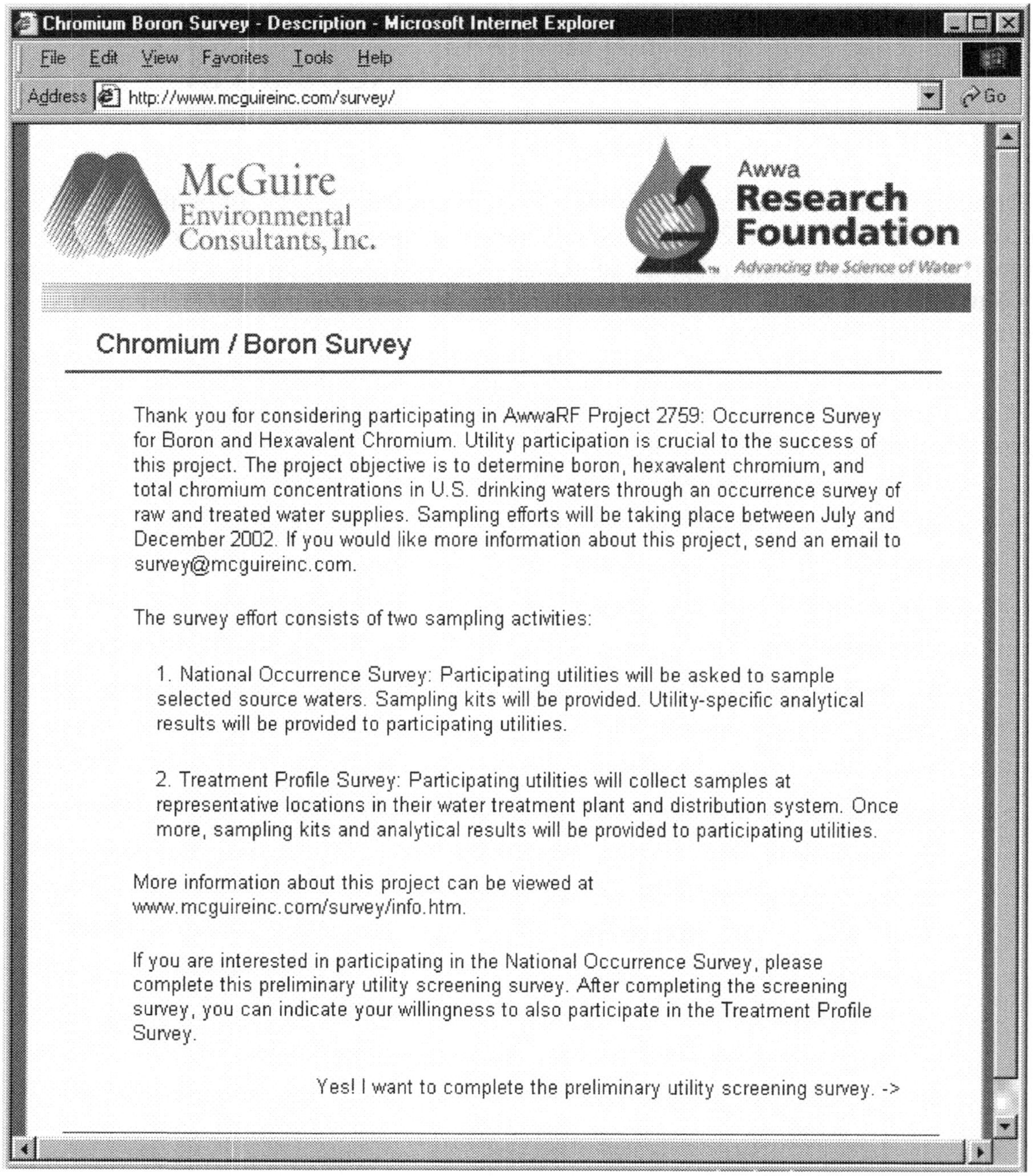

Figure 4-1. Online utility screening survey

McGuire Environmental Consultants, Inc.

Chromium / Boron Survey

Thank you for considering participating in AwwaRF Project 2759: *Occurrence Survey for Boron and Hexavalent Chromium*. Utility participation is crucial to the success of this project. The project objective is to determine boron, hexavalent chromium, and total chromium concentrations in U.S. drinking waters through an occurrence survey of raw and treated water supplies. Sampling efforts will be taking place between July and December 2002. If you would like more information about this project, project summary information can be viewed at:

www.mcguireinc.com/survey/info.htm
or send an email to: survey@mcguireinc.com.good

The survey effort consists of two sampling activities:

1. National Occurrence Survey: Participating utilities will be asked to sample selected source waters. Sampling kits will be provided. Utility-specific analytical results will be provided to participating utilities.

2. Treatment Profile Survey: Participating utilities will collect samples at representative locations in their water treatment plant and distribution system. Once more, sampling kits and analytical results will be provided to participating utilities.

If you are interested in participating in the National Occurrence Survey, please complete this preliminary utility screening survey. After completing the screening survey, you can indicate your willingness to also participate in the Treatment Profile Survey.

Please provide your contact information. Sample kits will be delivered to the Sampling Coordinator address.

Utility Name ________________

Your Name ________________

Email Address ________________

Phone Number ________________

☐ Check here if you are the Sampling Coordinator

Sampling Coordinator Information

Name ________________

Address ________________

City / State / Zip ________________

Email Address ________________

Phone Number ________________

What is the average daily production of your utility? ________________ MGD

Please provide the following information for your primary water sources. Up to five sources can be listed.

		Source 1	Source 2	Source 3	Source 4	Source 5
Source Name						
Water Type	Ground	☐	☐	☐	☐	☐
	Surface	☐	☐	☐	☐	☐
Daily Production (MGD)						
Treatment Process	Conventional + Alum	☐	☐	☐	☐	☐
	Conventional + Ferric	☐	☐	☐	☐	☐
	ILF/DF + Alum	☐	☐	☐	☐	☐
	ILF/DF = Ferric	☐	☐	☐	☐	☐
	Softening	☐	☐	☐	☐	☐
	FE/MN	☐	☐	☐	☐	☐
	MF/UF	☐	☐	☐	☐	☐
	NF/RO	☐	☐	☐	☐	☐
	Disinfection Only	☐	☐	☐	☐	☐
	None	☐	☐	☐	☐	☐
Disinfection Type (Plant)	Free Chlorine	☐	☐	☐	☐	☐
	Combined Chlorine	☐	☐	☐	☐	☐
	UV	☐	☐	☐	☐	☐
	Ozone	☐	☐	☐	☐	☐
	Chlorine Dioxide	☐	☐	☐	☐	☐
	None	☐	☐	☐	☐	☐
Disinfection Type (Distribution System)	Free Chlorine	☐	☐	☐	☐	☐
	Combined Chlorine	☐	☐	☐	☐	☐
	None	☐	☐	☐	☐	☐
Total Chromium Concentration	< 25 µg/L	☐	☐	☐	☐	☐
	25-50 µg/L	☐	☐	☐	☐	☐
	> 50 µg/L	☐	☐	☐	☐	☐
Total Boron Concentration	<0.5 mg/L	☐	☐	☐	☐	☐
	0.5-1 mg/L	☐	☐	☐	☐	☐
	>1 mg/L	☐	☐	☐	☐	☐

Figure 4-2. Paper utility screening survey

The project team mailed printed screening surveys to 1275 drinking water utilities described in the AWWA Water:\Stats database with instructions for completing either the online or print survey. The screening surveys were designed to collect the information necessary for characterizing each participating utility and up to five source waters to be sampled. The following categories of information were collected:

- Contact Information
- Utility Average Daily Production (MGD)
- Source Water Information (up to 5 source waters may be listed)
    - Source Name
    - Source Water Type
        - Ground
        - Surface
    - Daily Production (MGD)
    - Treatment Type
        - None = No treatment or disinfection
        - Disinfection Only = Disinfection only without additional treatment process
        - Conventional + Alum = Conventional flocculation, sedimentation, and filtration using an alum coagulant
        - Conventional + Ferric = Conventional flocculation, sedimentation, and filtration using an iron coagulant
        - ILF/DF + Alum = Inline or direct filtration using an alum coagulant
        - ILF/DF + Ferric = Inline or direct filtration using an iron coagulant
        - Fe/Mn = Iron and manganese control
        - Softening = Water softening treatment processes
        - MF/UF = Microfiltration or ultrafiltration membranes
        - NF/RO = Nanofiltration or reverse osmosis membranes
    - Disinfection Type (Plant)
        - Ozone
        - Chlorine Dioxide
        - Free Chlorine
        - Combined Chlorine
        - UV
        - None

- Disinfection Type (Distribution System)
  - Free Chlorine
  - Combined Chlorine
  - None
- Total Chromium Concentration Estimate
- Total Boron Concentration Estimate

The first phase of active utility recruitment using online and printed screening surveys brought the total number of participating drinking water utilities to 148. These water utilities represented 325 source waters. While this was a successful effort, some geographical regions of the country remained underrepresented.

The second phase of utility recruitment using direct contact by phone targeted those geographical regions of the country that were previously underrepresented. This recruiting effort brought the total number to 189 participating utilities representing 407 source waters.

CHARACTERIZATION OF PARTICIPATING UTILITIES

The population of participating utilities included in the NCBOS sampling effort will be described in this section by their respective source waters including their location and type. Figure 4-3 displays the participating utility locations and the source water types that they agreed to sample.

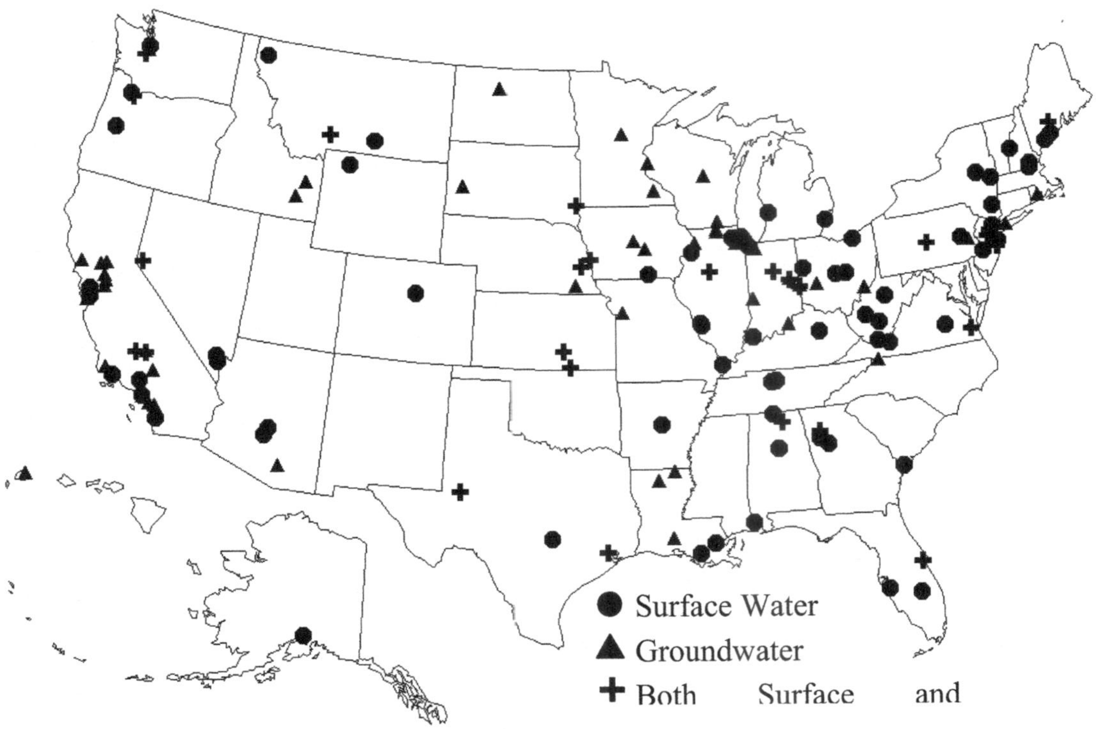

Figure 4-3. Participating utilities and source water types to be sampled

The participating utilities are distributed across 41 states, with the greatest number of utilities located in California, followed by Illinois and Indiana. Of the 407 source waters recruited to be sampled, 273 (67%) were groundwater sources and 134 (33%) were surface water sources. This can be compared with the USEPA FY2003 Public Water System Inventory Data (USEPA, 2004) where 78% of community water systems are supplied by groundwater and 22% are supplied by surface water. Table 4-1 describes the number of recruited utilities and their source waters by source type and state.

Table 4-1
Recruited utilities and source waters by state

State	Utilities	Total Sources	Groundwater Sources	Surface Water Sources
AK	1	1	0	1
AL	4	11	3	8
AR	1	2	0	2
AZ	3	4	1	3
CA	41	118	105	13
CO	1	1	0	1
CT	0	0	0	0
DE	0	0	0	0
FL	4	6	3	3
GA	3	7	1	6
HI	1	1	1	0
IA	5	10	8	2
ID	2	6	6	0
IL	19	31	23	8
IN	18	40	32	8
KS	2	5	3	2
KY	1	2	0	2
LA	5	11	7	4
MA	0	0	0	0
MD	0	0	0	0
ME	3	5	2	3
MI	2	2	0	2
MN	2	6	6	0
MO	2	3	1	2
MS	0	0	0	0
MT	4	9	3	6
NC	1	1	1	0
ND	1	1	1	0
NE	3	6	4	2
NH	3	3	0	3
NJ	6	15	6	9
NM	0	0	0	0
NV	3	7	4	3
NY	4	4	1	3
OH	6	8	4	4
OK	0	0	0	0
OR	3	4	1	3
PA	5	14	12	2
RI	1	1	1	0
SC	2	2	0	2
SD	3	9	8	1
TN	2	2	0	2
TX	3	6	2	4
UT	0	0	0	0
VA	3	6	1	5
VT	0	0	0	0
WA	3	11	8	3
WI	3	10	10	0
WV	8	10	0	10
WY	2	6	4	2
Totals	**189**	**407**	**273 (67%)**	**134 (33%)**

CHAPTER 5
IMPLEMENTATION OF THE NATIONAL OCCURRENCE SURVEY

PRELIMINARY SURVEY

Preliminary survey sample kits were set to 20 participating utilities that had committed to participate in the project with in-kind support at the time of the proposal submittal. This preliminary survey was intended to assist the methods development effort of the project, streamline the sample kit preparation, delivery, and receipt and data management of sample information and results.

The preliminary survey sample kits included 8 sample bottles. Each bottle represented a different preservative and sample handling approach –in both the field and analytical laboratory. Appendix B describes each of the preliminary survey sample bottles in detail; and it contains the cover letter, instructions, and survey questionnaire used with the preliminary survey.

All 20 of the preliminary survey sample kits were returned. The results from each of the different sample bottles resulted in the final sample kit which used two bottles for the subsequent regular survey.

REGULAR SURVEY

Building upon the experience gained from the preliminary survey, the sample kit design was finalized as described in Chapter 2. The sample kit letter, instructions, and questionnaire were also refined. Appendix C contains these materials included in the regular survey sample kits. Figure 5-1 shows the sample kit preparation for the regular survey.

Figure 5-1. Sample kit preparation

The regular survey was conducted in two rounds as defined by the utility recruitment effort described in Chapter 4. During the Round One sampling effort, over 300 sample kits were sent to participating utilities during May, June, and July 2003; 85 sample kits were sent to participating utilities added after Round One sampling in December 2003 for the Round Two sampling effort.

Of the 407 source water sample kits sent to participating utilities in both the preliminary and regular surveys, 342 sample kits were returned and analyzed.

CHAPTER 6
NATIONAL OCCURRENCE OF CHROMIUM AND BORON

GENERAL OCCURRENCE RESULTS

Table 6-1 describes the number of total sample results and non-detect results for boron, total chromium, and hexavalent chromium.

Table 6-1
General sample detection results

Parameter	Total Sample Results	MDL (μg/L)	Non-detect Results	% of Non-detect Results
Boron	341	2.0	4	1.2%
Total Chromium	342	0.6	133	38.8%
Cr(VI)	341	0.2	195	57.2%

It is interesting to note that over 57% of the hexavalent chromium sample results were below the MDL of 0.2 μg/L, while only 38.8% of the total chromium results were below the MDL of 0.6 μg/L. To further illustrate the survey sample results for boron, total and hexavalent chromium, Figures 6-1 and 6-2 display the cumulative distribution of the boron sample results and the total and hexavalent chromium sample results from the NCBOS, respectively. A clear difference can be seen between the populations of total chromium results and hexavalent chromium results in Figure 6-2.

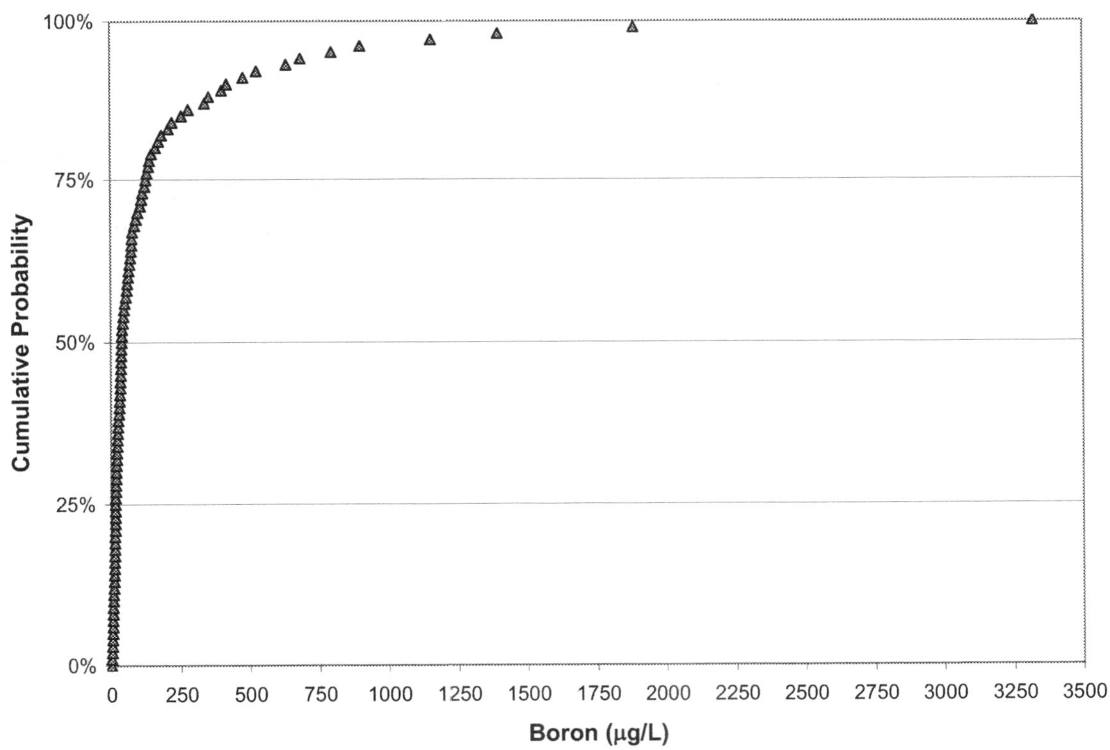

Figure 6-1. Cumulative distribution of NCBOS boron occurrence data

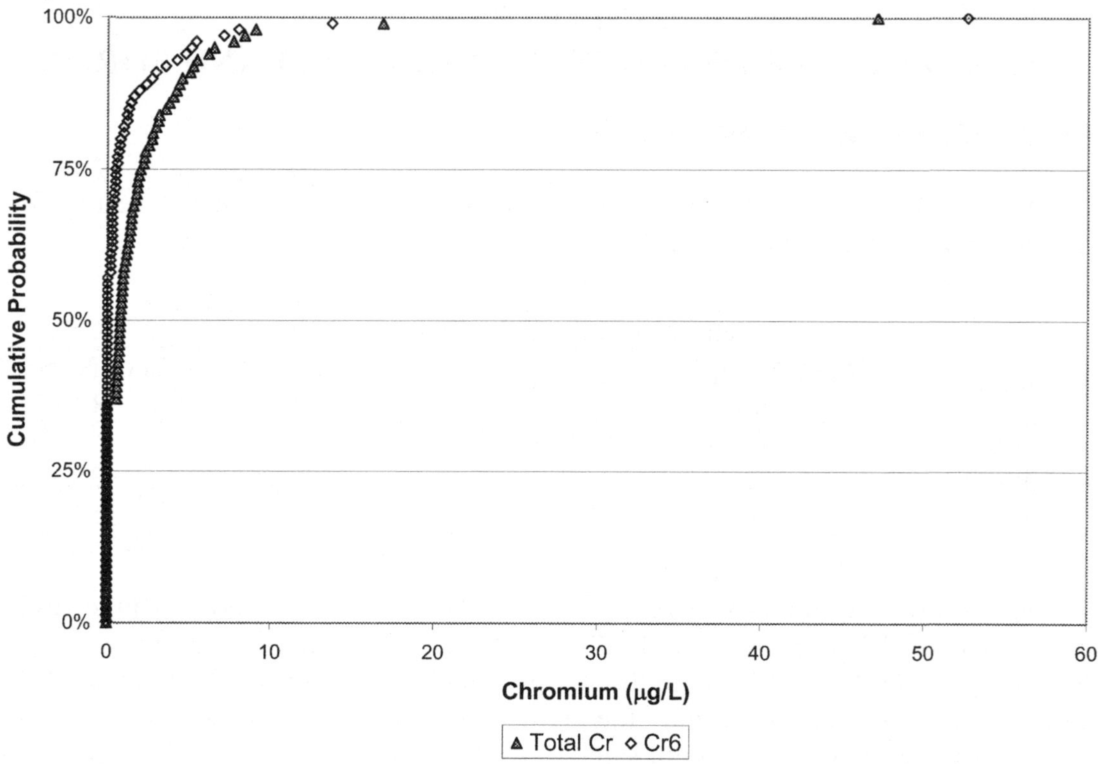

Figure 6-2. Cumulative distributions of NCBOS total and hexavalent chromium occurrence data

Source water type is an important characteristic for categorizing the sample results. Figures 6-3, 6-4, and 6-5 show box-and-whisker plots for surface water and groundwater sample results for boron, total chromium, and hexavalent chromium. All of the box-and-whisker plots presented in this chapter display the number of observations, or count, included in the data population being described. In Figure 6-3 for example, the groundwater boron population included 228 results, while the surface water boron population included 113 results.

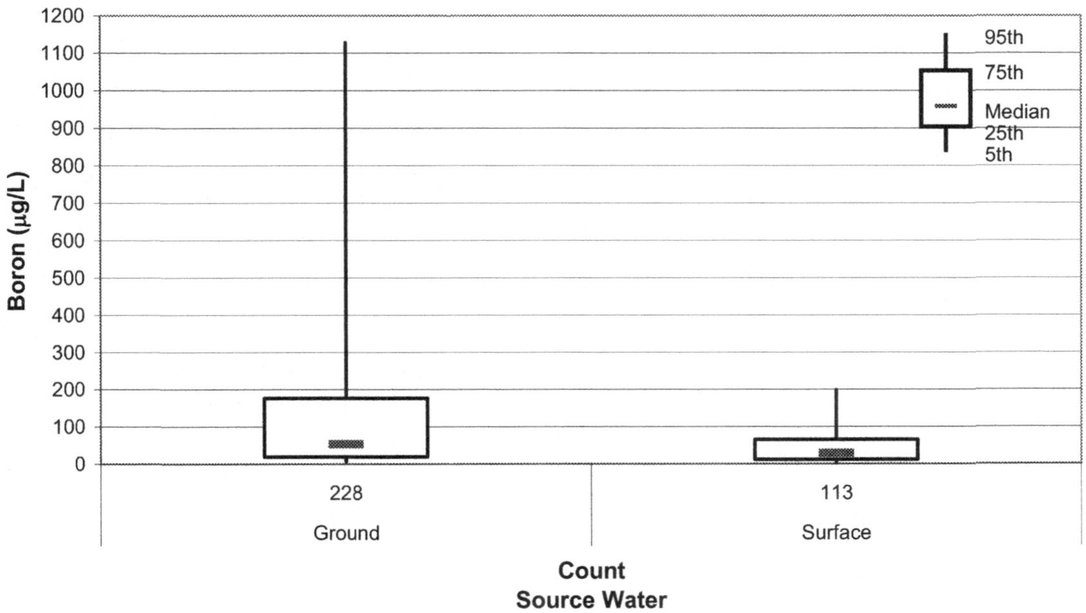

Figure 6-3. NCBOS boron data by source water type

While the median boron concentrations for boron in groundwaters and surface waters are similar (groundwater boron median = 54.1 µg/L; surface water boron median = 29.0 µg/L), the groundwater boron data is appears to be distributed in a log-normal fashion. The 75th and 95th percentile values for the groundwater data are significantly higher than the respective values for surface water boron data.

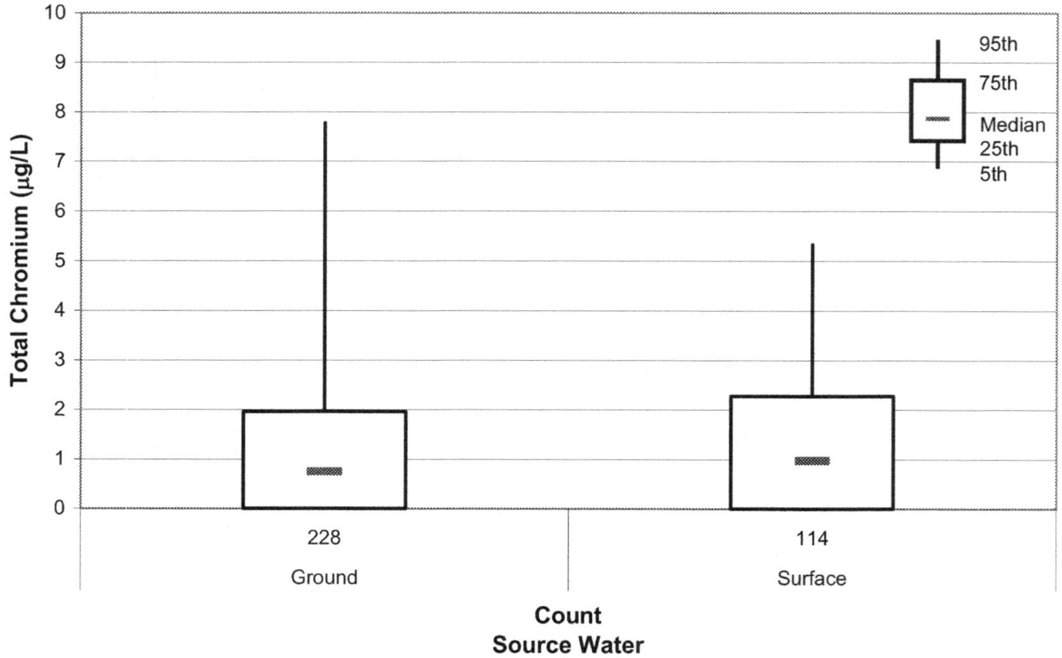

Figure 6-4. NCBOS total chromium data by source water type

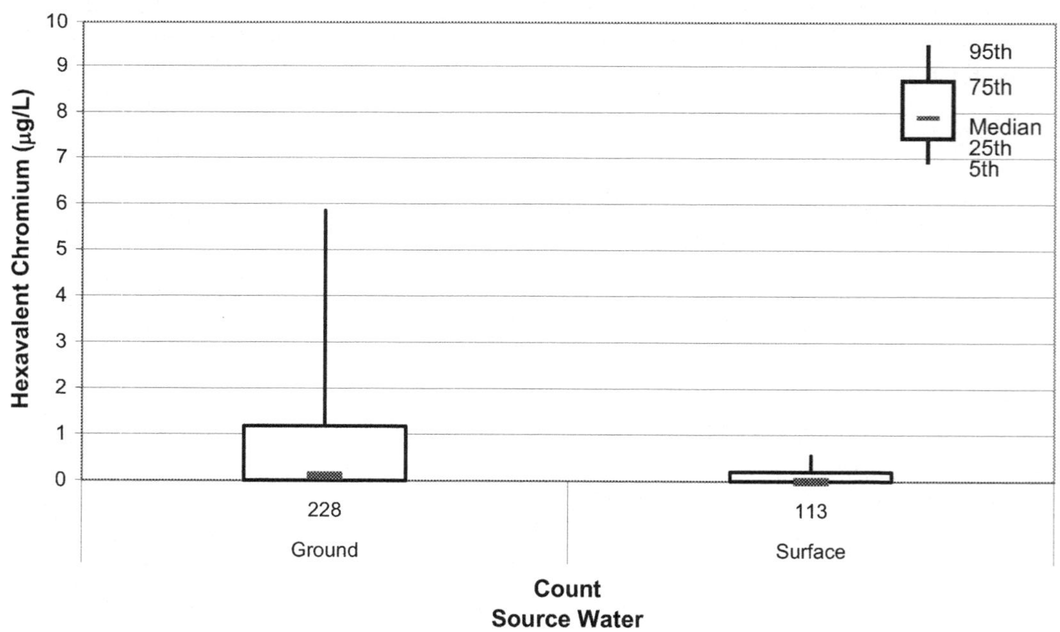

Figure 6-5. NCBOS hexavalent chromium data by source water type

The box-and-whisker plots for total and hexavalent chromium by source water show that while total chromium is found at approximately equal concentrations in both groundwaters and surface waters, hexavalent chromium is not found in surface waters to nearly the same degree as in ground waters. This speciation trend by source type is illustrated in Figure 6-6 where paired total and hexavalent chromium concentration are plotted for both groundwaters and surface waters. This graph only shows those results below 15 μg/L.

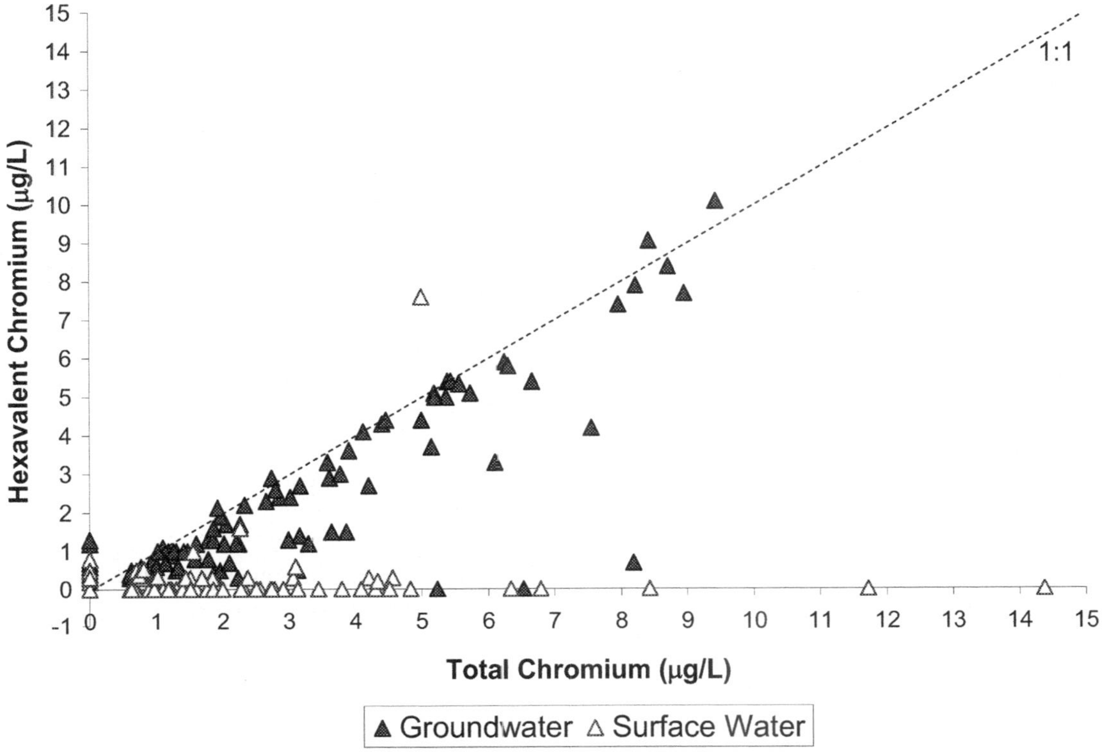

Figure 6-6. Paired total and hexavalent chromium results (below 15 μg/L) by source type

As illustrated in Figure 6-6, total chromium concentrations in surface waters are – with few exceptions – made up predominantly by trivalent chromium. This figure also shows the ability of the project team's analytical methods to adequately capture total chromium concentrations in all source waters since no sample pairs reside significantly above the 1:1 line.

Another important view of the occurrence data is by system size category. Population served data was obtained from the USEPA Safe Drinking Water Information System (SDWIS) for each participating utility. Three system size categories were defined using the population served data for each public water system: <10,000; 10,000 – 50,000; and >50,000. Figures 6-7, 6-8, and 6-9 display box and whisker plots of boron, total chromium, and hexavalent chromium by system size category.

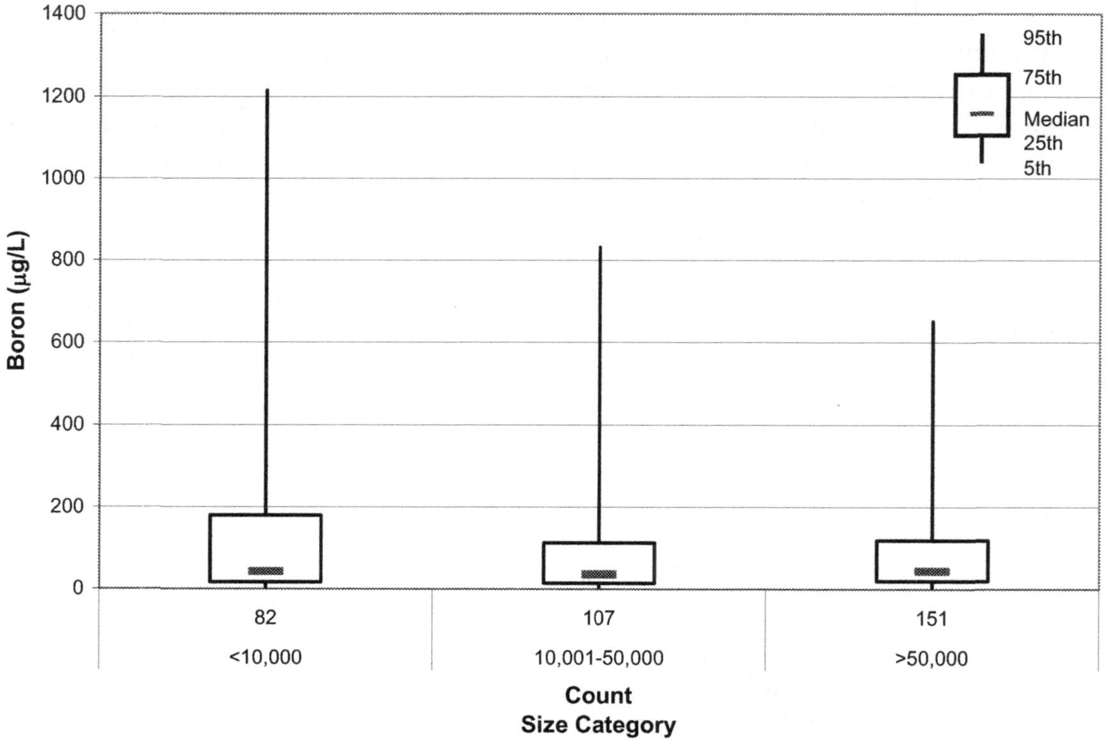

Figure 6-7. NCBOS boron data by system size category

Boron occurrence is fairly consistent by system size category, with only the 95th percentile (and 75th percentile in the case of the <10,000 size category) increasing with decreasing system size.

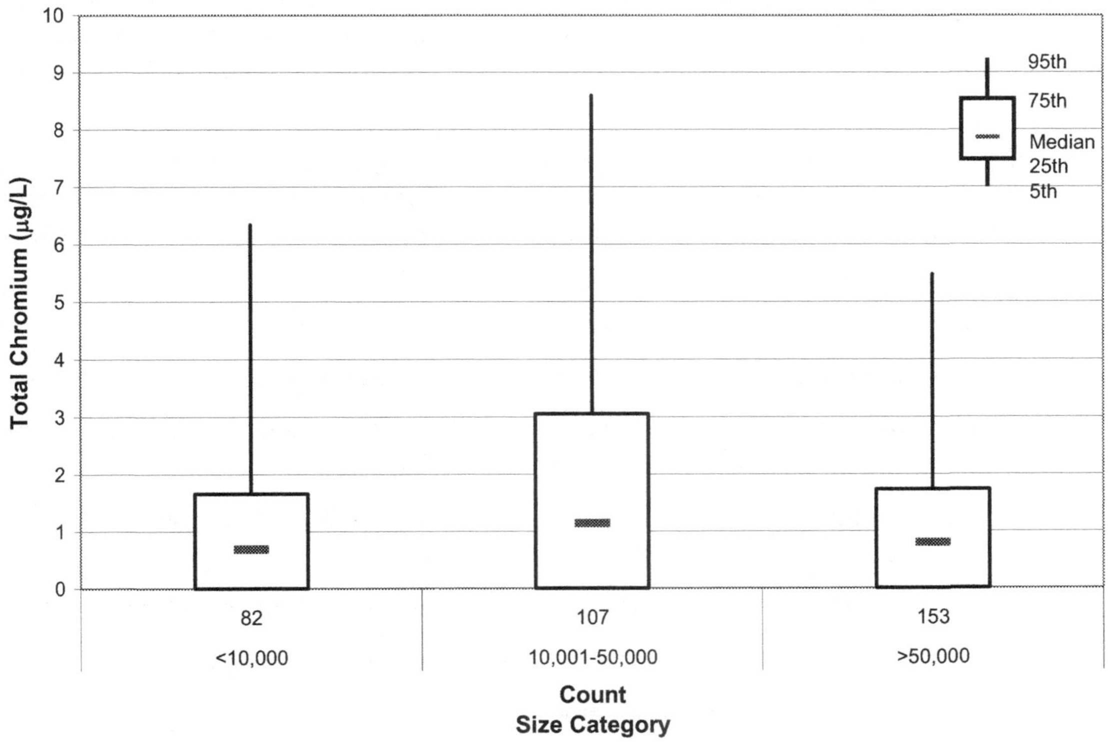

Figure 6-8. NCBOS total chromium data by system size category

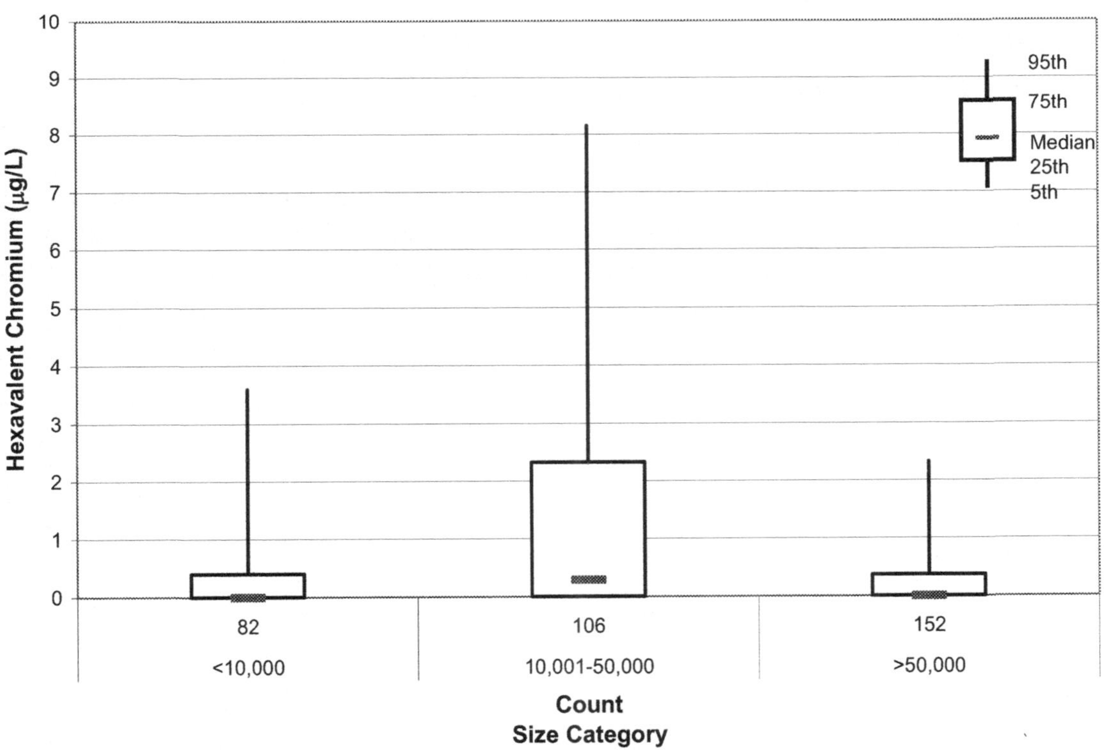

Figure 6-9. NCBOS hexavalent chromium data by system size category

Both total and hexavalent chromium occurrence data sets exhibited elevated concentrations in the medium 10,001 – 50,000 size category. This may be an artifact attributed primarily to source water type since the medium size category is dominated by groundwater sources.

Finally, the occurrence data can be described by the treatment type used to treat each source water sampled. This view of the source water occurrence data is relevant to the treatment profile results in next chapter where boron and chromium removal by various treatment types is discussed. The treatment types included in this evaluation are:

- **None** = No treatment or disinfection
- **Disinfection Only** = Disinfection only without addition treatment processes
- **Conventional + Alum** = Conventional flocculation, sedimentation, and filtration using an alum coagulant
- **Conventional + Ferric** = Conventional flocculation, sedimentation, and filtration using an iron coagulant
- **ILF/DF + Alum** = Inline or direct filtration using an alum coagulant
- **ILF/DF + Ferric** = Inline or direct filtration using an iron coagulant
- **Fe/Mn** = Iron and manganese removal
- **Softening** = Water softening treatment processes
- **GAC** = Granular activated carbon
- **MF/UF** = Microfiltration or ultrafiltration membranes
- **NF/RO** = Nanofiltration or reverse osmosis membranes

Figures 6-10, 6-11, and 6-12 describe boron, total chromium, and hexavalent chromium occurrence by treatment type. It is important to note the number of results represented by each treatment type in the following figures. Some of the treatment types are limited to 2 or 3 results; this limits the significance of these results.

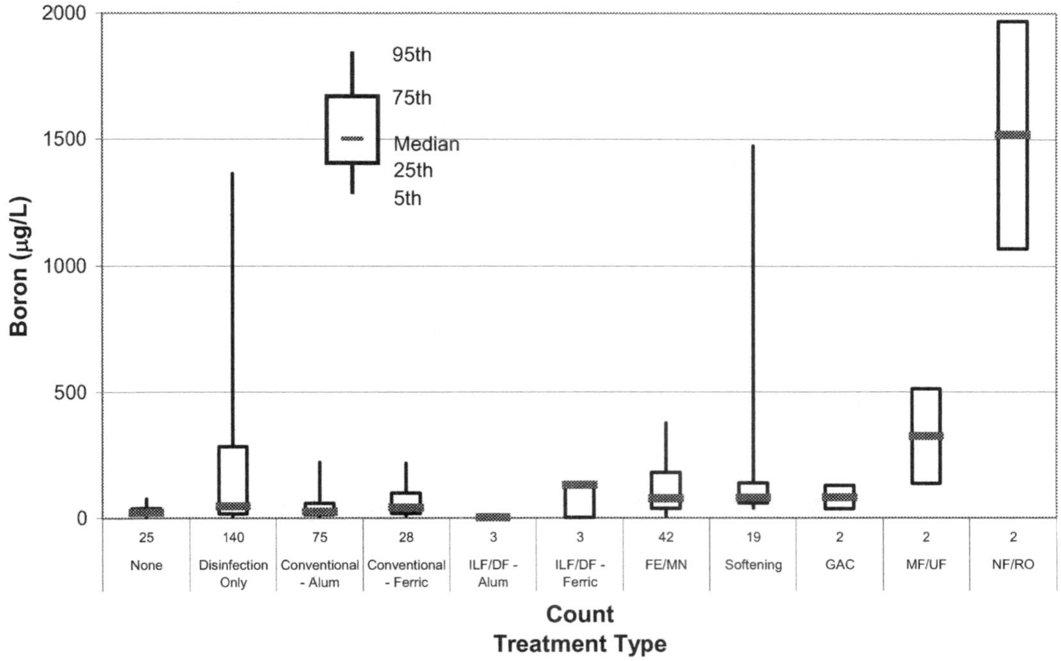

Figure 6-10. NCBOS boron data by treatment type

The boron occurrence data displayed in Figure 6-10 indicate the highest boron concentrations are found in source waters treated by NF/RO membranes, followed by source waters treated by MF/UF, softening, and disinfection only.

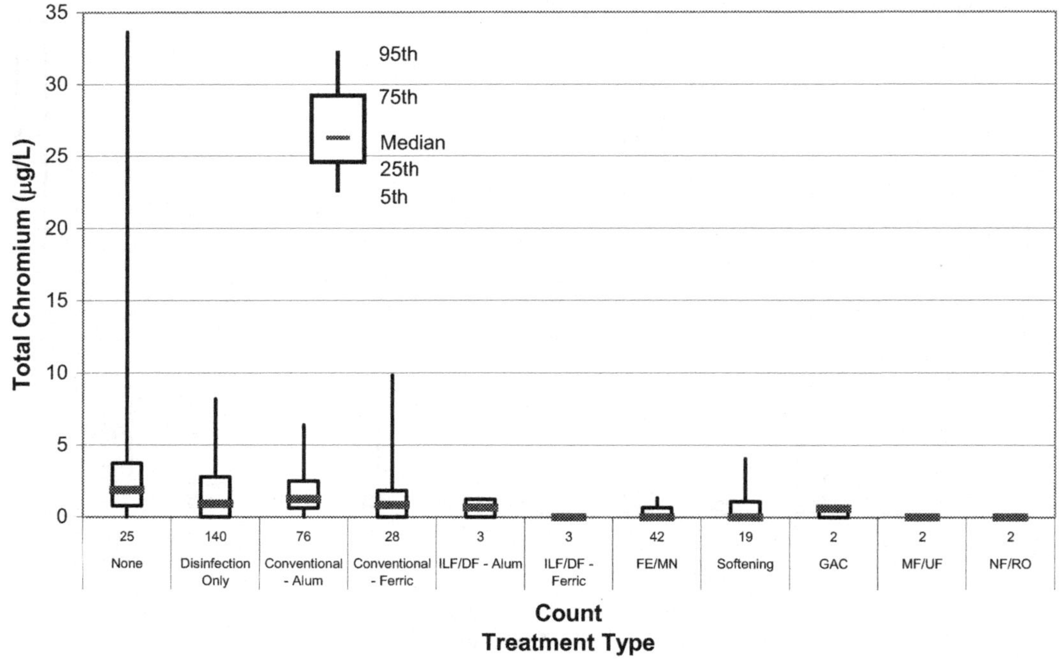

Figure 6-11. NCBOS total chromium data by treatment type

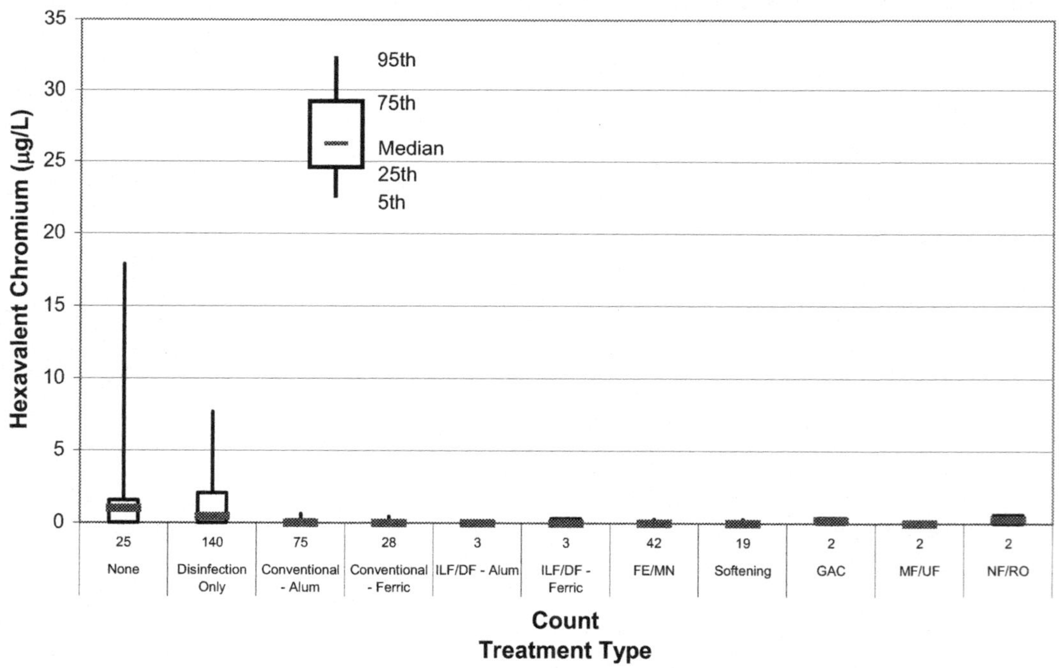

Figure 6-12. NCBOS hexavalent chromium data by treatment type

The occurrence trends for total and hexavalent chromium are essentially the same for all treatment types except conventional treatment where total chromium concentrations are higher. Since these treatment types are typically associated with surface waters, this result is consistent with the chromium speciation occurrence trends by source water presented previously in Figure 6-6. This conclusion can also be made for waters treated by softening. The highest chromium concentrations are found in source waters not treated or just disinfected – typically associated with groundwaters.

REGIONAL SIGNIFICANCE OF OCCURRENCE RESULTS

Geographic trends are also important when characterizing the NCBOS data. For this analysis, EPA regions serve as the basis for combining states into regions. Figure 6-13 depicts the EPA region boundaries.

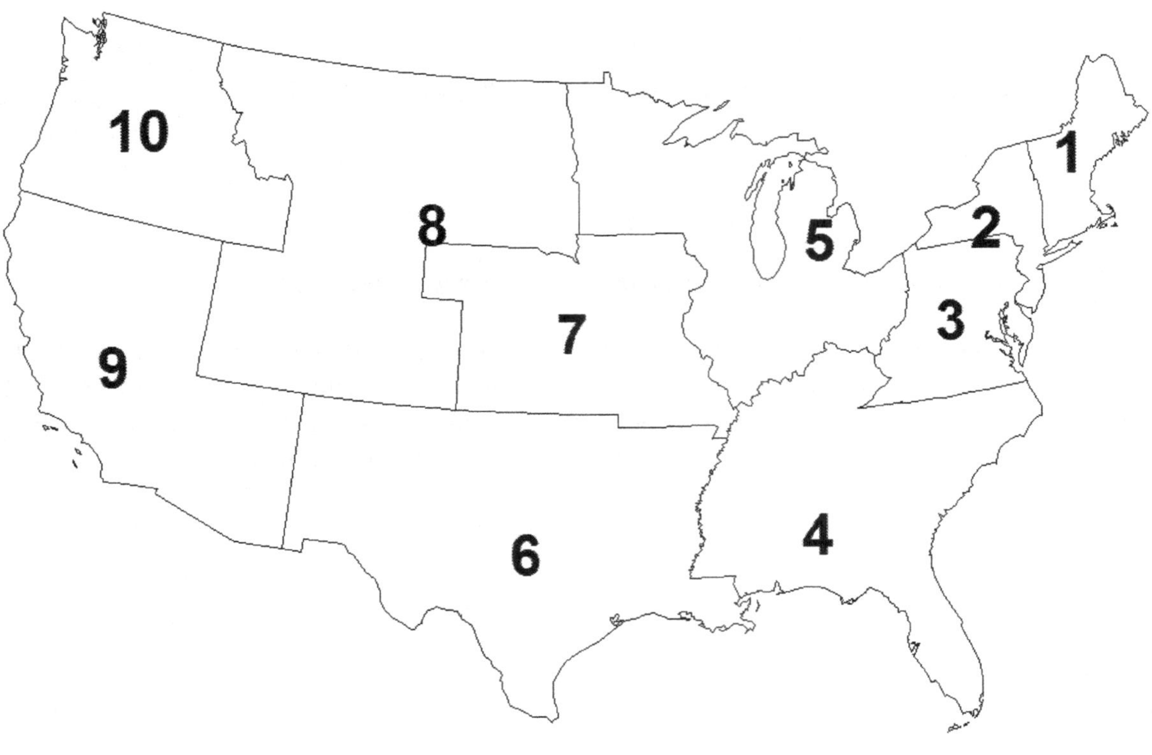

Figure 6-13. EPA region boundaries

The following box-and-whisker plots show boron, total chromium, and hexavalent chromium by EPA region and source water type.

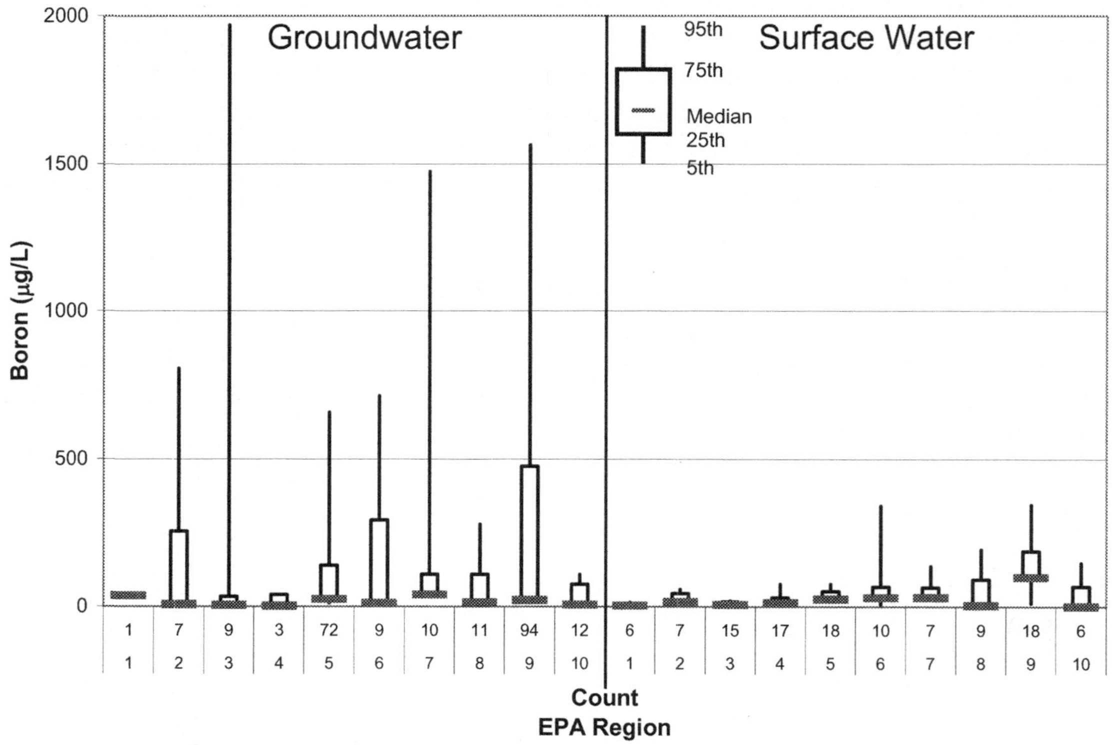

Figure 6-14. NCBOS boron data by EPA region and source water type

The boron data presented in Figure 6-14 indicates groundwater occurrence is more variable by region than surface water occurrence. No clear occurrence patterns exist by EPA region for groundwater boron since those concentrations are typically low; however, surface water boron occurs to a greater degree in the western U.S than in the eastern U.S.

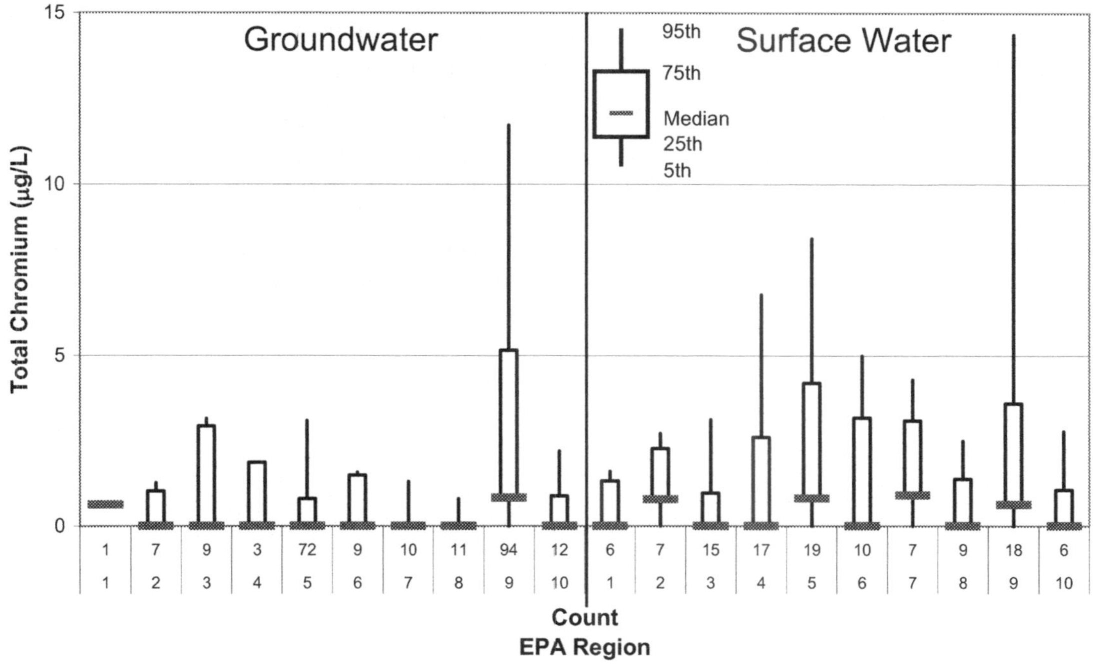

Figure 6-15. NCBOS total chromium data by EPA region and source water type

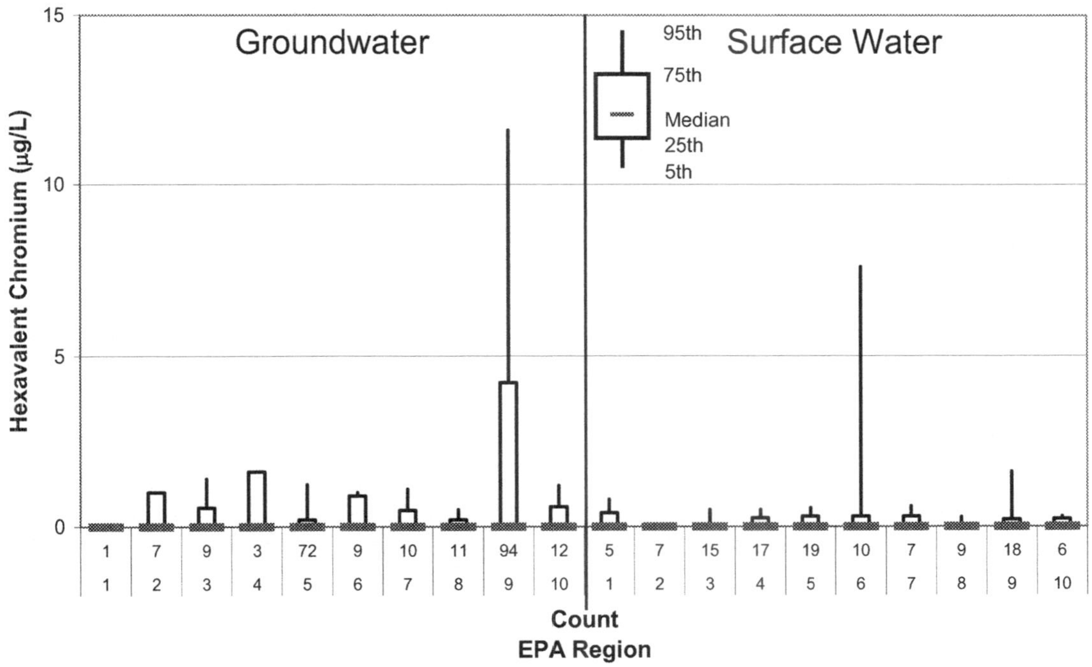

Figure 6-16. NCBOS hexavalent chromium data by EPA region and source water type

The data presented in Figures 6-15 and 6-16 show that chromium occurrence particularly in groundwaters – both total and hexavalent – is greatest in EPA Region 9 which includes California. This trend may be biased due to the participation of California water utilities in this project that were already aware of chromium presence in their source waters. This trend may also be representative of the true regional occurrence patterns.

Total chromium occurrence is more variable by region in surface waters than groundwaters. No clear regional patterns exist for total chromium in surface waters. In groundwaters, total chromium exists to the greatest degree in EPA Region 9.

These data again confirm the speciation trend by source type described in Figure 6-6 where surface water chromium tends to exist predominantly as trivalent chromium.

COMPARISON OF NCBOS DATA WITH MEANINGFUL OCCURRENCE DATA SOURCES

The NCBOS data is compared with data from the meaningful occurrence data sources identified in the project to determine if the other meaningful data sources can be used for assessing drinking water occurrence of boron, total chromium, and hexavalent chromium. The following figures use cumulative probability distribution graphs to display similarities and differences in the occurrence data sets by parameter and source type. Figure 6-17 shows the NCBOS boron occurrence data to be lower in general than the NWIS-I data when comparing respective source waters. These results suggest the NWIS-I data which represent ambient source water samples may not be suitable for estimating boron occurrence in drinking water sources. When compared with the NIRS groundwater boron data, the NCBOS data tend to be higher. Finally, the NCBOS groundwater boron data are somewhat higher than the CalDHS groundwater boron data; while the NCBOS surface water boron data are lower than the CalDHS surface water boron data.

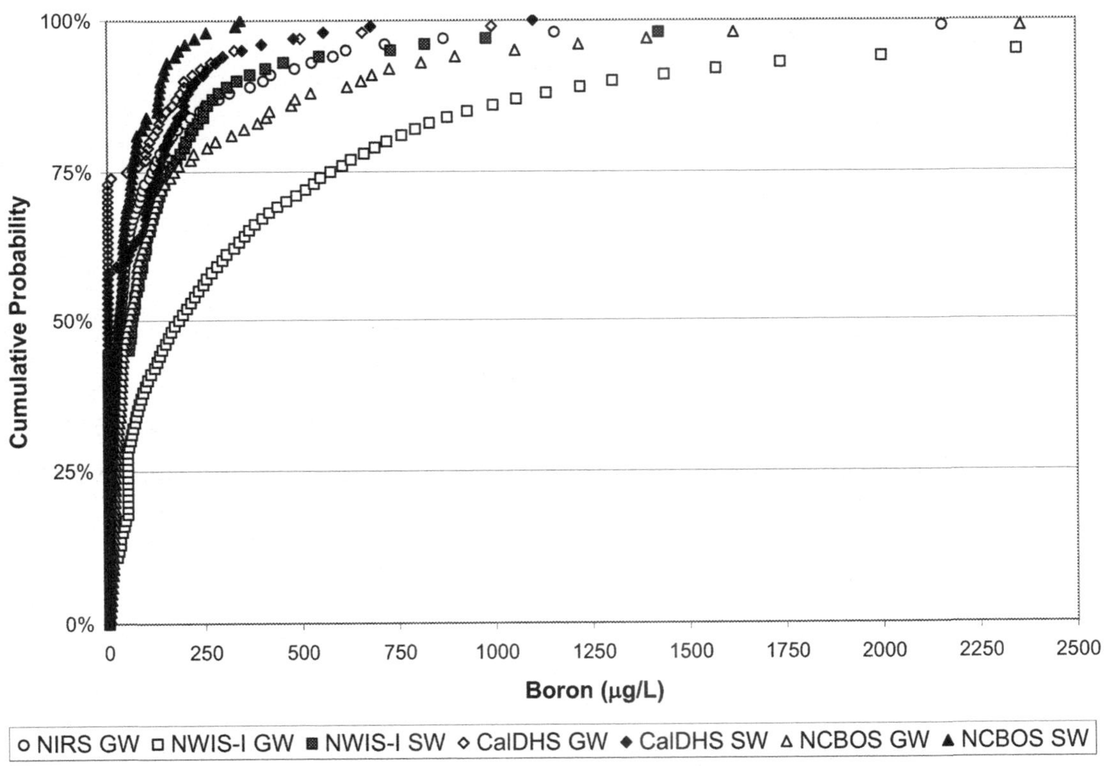

Figure 6-17. Comparison of boron occurrence by data set and source water

The NCBOS occurrence data for total chromium are quite different from the NIRS, CalDHS, and NWIS-I data. In Figure 6-18, the NCBOS data display a much higher degree of detectable chromium concentrations than the NIRS data. If the NCBOS data had been developed with the same higher detection limit as the NIRS (2 µg/L), 75% of the NCBOS groundwater total chromium results would have been below detection limit (BDL), while the 95% of the NIRS results are BDL. When compared with the CalDHS data, the NCBOS data again display a higher degree of detectable chromium concentrations. However, the upper percentile (>90% for groundwater and >95% for surface water) total chromium results from the CalDHS data set are than the corresponding NCBOS results. As previously discussed in Chapter 3 – because of its historical nature, poor detection limits, and sampling bias toward ambient waters unsuitable for drinking, the NWIS-I total chromium data appear unsuitable for estimating total chromium occurrence in drinking water sources.

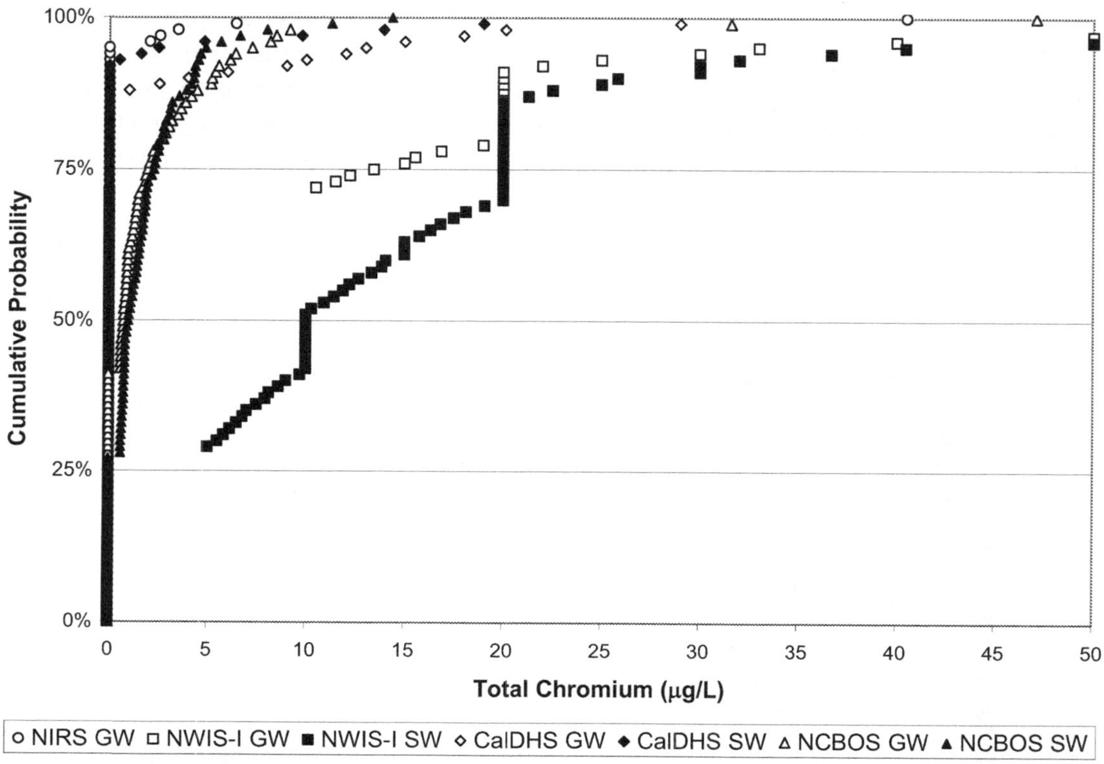

Figure 6-18. Comparison of total chromium occurrence by data set and source water

Finally, Figure 6-19 shows that the NCBOS surface water hexavalent chromium data correspond well with the CalDHS surface water data, but the CalDHS groundwater hexavalent chromium data are significantly higher than NCBOS groundwater data. The NCBOS groundwater hexavalent chromium data are fairly agreeable with the NWIS-I groundwater data, but NWIS-I surface water data tend to be higher than NCBOS surface water data. Given the issues previously discussed concerning the NWIS-I ambient source water data, the NCBOS data provides a new baseline for hexavalent chromium in drinking water sources.

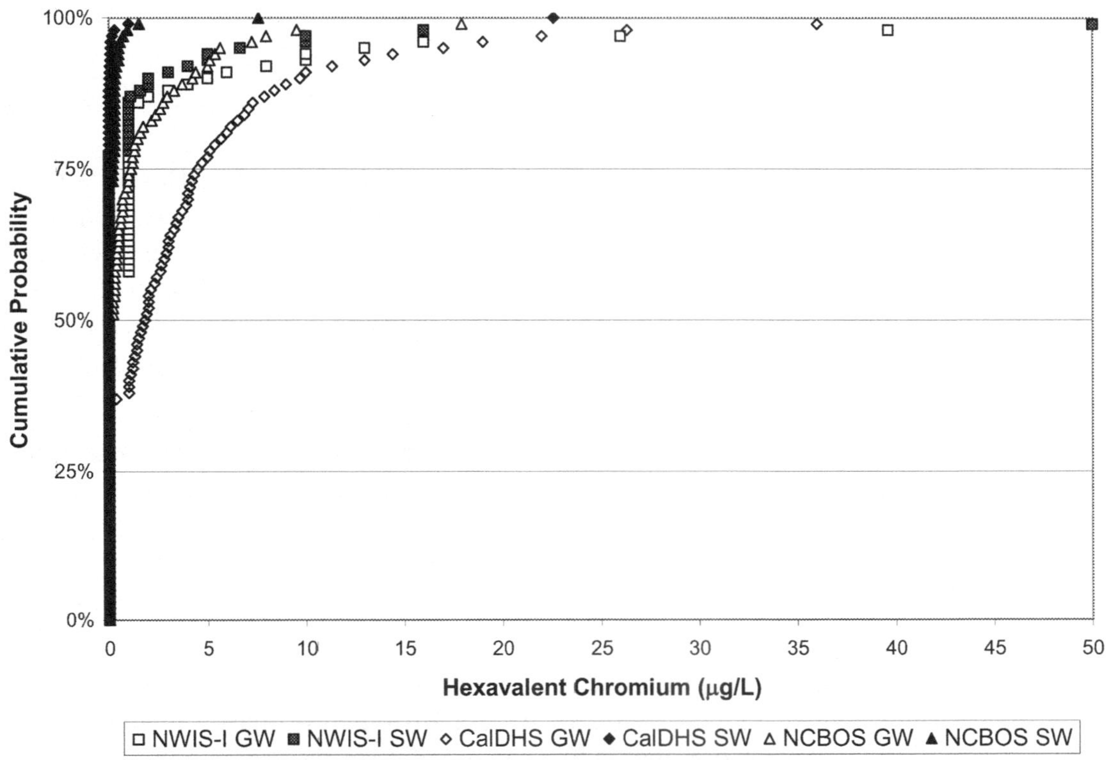

Figure 6-19. Comparison of hexavalent chromium occurrence by data set and source water

CONCLUSIONS

The occurrence results from the NCBOS are summarized in Table 6-2.

Table 6-2
NCBOS occurrence data summary

Parameter	Count (µg/L)	Average (µg/L)	Median (µg/L)	Maximum (µg/L)
Boron	341	167.9	40.0	3,321.3
Total Chromium	342	1.9	0.8	47.1
Hexavalent Chromium	341	1.1	0.0	52.6

Several important results were identified from the NCBOS data. Boron was found to occur higher in groundwater sources than surface water sources and occurred equally across water system size categories. Source waters treated by NF/RO membranes exhibited the highest boron occurrence by treatment type, followed by MF/UF, softening, and disinfection only. No clear occurrence patterns exist by EPA region for groundwater boron occurrence; however, surface water boron occurs to a greater degree in the western U.S than in the eastern U.S.

A majority of the hexavalent chromium sample results were found to be less than the MDL of 0.2 µg/L. Total chromium was found to occur equally in surface waters and groundwaters; however, hexavalent chromium is not found in surface waters to nearly the same degree as in groundwaters. NCBOS total chromium concentrations in surface waters are – with few exceptions – composed primarily of trivalent chromium. Several groundwater results show total chromium concentrations composed exclusively by hexavalent chromium. This speciation trend has not been previously reported.

Both total and hexavalent chromium exhibited elevated concentrations in the medium 10,001 – 50,000 size category. This may be an artifact attributed primarily to source water type since the medium size category is dominated by groundwater sources.

The occurrence trends for total and hexavalent chromium are essentially the same for all treatment types except conventional treatment where total chromium concentrations are higher. Since these treatment types are typically associated with surface waters, this result is consistent with the chromium speciation occurrence trends by source water previously reported. This conclusion can also be made for waters treated by softening. The highest chromium concentrations are found in source waters not treated or just disinfected – typically associated with groundwaters.

When evaluating regional occurrence trends, chromium – both total and hexavalent – is greatest in EPA Region 9 which includes California. Total chromium occurrence is more variable by region in surface waters than groundwaters. No clear regional patterns exist for total chromium in surface waters. However, regional patterns did exist for groundwater results where total chromium occurred to the greatest degree in EPA Region 9.

CHAPTER 7
NATIONAL TREATMENT PROFILES

As discussed in the previous chapter, hexavalent chromium and boron occur in US drinking water sources at levels that could have national significance for future regulatory concerns. The NCBOS data, however, was collected entirely for drinking water sources and did not include treated water sampling. As an initial screening for the potential impacts of these contaminants on drinking water utilities, source water monitoring information from this study identified the highest vulnerability for contaminant control requirements.

Profiling the occurrence and, in the case of chromium, shifts in speciation as a result of existing treatment technologies can provide direct insight into the likely transformation or removal of chromium and boron in US drinking water supplies from source to tap. Seventeen water utilities were selected from those that participated in the NCBOS for further evaluation in the treatment profile survey performed in this study (Table 7-1). Diagrams of the sampling locations for the participating utilities are contained in Appendix D. Seven utilities were selected to specifically assess the fate of chromium; another eight utilities were selected to specifically assess the fate of boron; and two utilities were selected to address the fate of both chromium and boron. The majority of the facilities included in the treatment profile study were groundwater supplies (12 of the 17 utilities), and among the groundwater systems, the technologies included in the profiling were (a) no treatment, (b) disinfection only, (c) iron/manganese removal, (d) granular activated carbon, and (e) nanofiltration membrane treatment.

Table 7-1
Descriptions of the treatment profile utilities

Source Type	Treatment Type	National Occurrence Survey Results: Cr(VI), µg/L	National Occurrence Survey Results: B, µg/L	Targeted for Treatment Profile t: Cr(VI)	Targeted for Treatment Profile t: B	Treatment Profile Sampling Locations: Source	Treatment Profile Sampling Locations: Clarified	Treatment Profile Sampling Locations: Treated	Treatment Profile Sampling Locations: Distribution System	Notes
Ground	Disinfection Only	52.6	855	Yes	Yes	X		X	X	Source represents multiple inter-connected wells
Ground	Disinfection Only	3.7	2500	No	Yes	X		X	X	
Ground	Disinfection Only	0.8	3320	No	Yes	X		X	X	
Ground	Disinfection Only	15.3	60.4	Yes	No	X		X	X	
Ground	Disinfection Only			Yes	No	X		X	X	
Ground	Fe/Mn	0	673	No	Yes	X	X	X	X	
Ground	Fe/Mn	0.6	354	Yes	Yes	X	X	X	X	Source is a blend of two collector wells; intermediate process sample is blend of other well sources with Fe/Mn treated source water.
Ground	Ferric	0	163	No	Yes	X	X	X	X	
Ground	GAC	5.4	30.8	Yes	No	X		X	X	
Ground	GAC		128	Yes	No	X	X	X	X	Raw water is a blend of multiple wells; intermediate sample is post aeration
Ground	NF/RO			No	Yes	X		X	X	Nanofiltration of brackish groundwater
Ground	None	18.6	68.5	Yes	No	X			X	
Surface	Alum	0	345	No	Yes	X	X	X	X	
Surface	Alum	1	147	Yes	No	X	X	X	X	
Surface	Alum	0.3	343	No	Yes	X	X	X	X	
Surface	Alum	7.6	37.5	Yes	No	X	X	X	X	
Surface	Ferric		238	No	Yes	X	X	X	X	Facility applies ozone between source and clarified water sampling locations

For all systems where any treatment is indicated, disinfection was also practiced at some point prior to the distribution system sampling location (Table 7-2). The presence of oxidants – such as free and combined chlorine – can affect the speciation of chromium in drinking water (Brandhuber et al., 2004; Clifford and Chau 1987). The shift in chromium species is kinetically dependent on the oxidant applied. Prior research has shown that free chlorine creates substantial shifts from Cr(III) to Cr(VI) in less than one hour, while measurable but lower intensity shifts occur after days of contact time with chloramines. The treatment profile sampling was designed to reflect a range of contact times and oxidant types to assess chromium speciation impacts.

Table 7-2
Oxidation and disinfection processes at treatment profile utilities

Source Type	Treatment Type	First Point of Oxidant Addition: Location	First Point of Oxidant Addition: Type	Treated Water Disinfectant	Distribution System Disinfectant
Ground	Disinfection Only		None Indicated		Free Chlorine
	Disinfection Only	Clearwell	Free Chlorine	Free Chlorine	Free Chlorine
	Disinfection Only	Pipeline from Well	Free Chlorine	Free Chlorine	Free Chlorine
	Disinfection Only				Free Chlorine
	Disinfection Only	Pipeline from Well	Free Chlorine	Free Chlorine	Free Chlorine
	Fe/Mn	Post-Aeration	Free Chlorine, Potassium Permanganate	Free Chlorine	Chloramines
	Fe/Mn	Pre-filtration	Free Chlorine	Chloramines	Chloramines
	Fe/Mn	Pre-filtration	Free Chlorine	Free Chlorine	Free Chlorine
	GAC		None Indicated		Free Chlorine
	GAC	Clearwell	Free Chlorine	Free Chlorine	Chloramines
	NF/RO		None Indicated		Chloramines
	None		No Disinfectant or Oxidant Applied		
Surface	Alum	Clearwell	Free Chlorine	Free Chlorine	Free Chlorine
	Alum	Clearwell	Free Chlorine	Free Chlorine	Free Chlorine
	Alum	Rapid Mix	Free Chlorine	Free Chlorine	Free Chlorine
	Alum	Clearwell	Free Chlorine	Free Chlorine	Chloramines
	Ferric	Pretreatment	Ozone	Free Chlorine	Free Chlorine

CHROMIUM TREATMENT PROFILE RESULTS

The analysis of total chromium and hexavalent chromium for treatment profile systems was performed for all participating systems, including those targeted for only boron treatment profiling. Thus, in several cases, very low levels and even non-detect results occurred in the source water samples for the profiled systems. Those utilities included in the study specifically to capture chromium occurrence results, however, are noted throughout this section and the discussion of findings is focused on the behaviors observed for these systems in particular.

The treatment profile survey samples were collected approximately six-months following the original sample collection performed as part of the national survey. As demonstrated in Table 7-3, variations in source water chromium concentrations were observed between the two sampling events. For surface water facilities, temporal variability is an expected outcome. Groundwater sources, typically considered less variable, also were sampled in many cases from sources that represented blends of multiple well supplies. The variability shown in the groundwater sources, therefore, is likely to be representative of the impact of the operational practices by water utilities using manifolded well supplies. Only in one case were the treatment profile samples in the source water near or below detection for those sites targeted for Cr(VI) profiling. In all other cases, the chromium levels were well above detection, permitting determination of the fate of Cr(VI) and total chromium through the distribution system.

The occurrence of Cr(VI) relative to total chromium remained relatively consistent between the two source water sampling events even though the magnitude of Cr(VI) and total chromium varied (Table 7-3). Hexavalent chromium was the dominant specie of chromium in the potable water sources sampled during the treatment profile survey.

Table 7-3
Comparison of survey and treatment profile source water chromium levels

Source Type	Treatment Type	National Occurrence Survey Source Chromium, μg/L		Treatment Profile Source Chromium, μg/L		Treatment Profile Targeted Cr(VI)
		Cr(VI)	Total Cr	Cr(VI)	Total Cr	
Ground Water	Disinfection Only	52.6	47.1	22.2	31.2	Yes
	Disinfection Only	3.7	5.2	1.3	2.2	No
	Disinfection Only	0.8	1	0.7	1.4	No
	Disinfection Only	9.07	8.42	14.0	13.2	Yes
	Fe/Mn	< 0.2	< 0.6	< 0.2	< 0.6	No
	Fe/Mn	0.6	0.8	< 0.2	< 0.6	Yes
	Fe/Mn	< 0.2	< 0.6	< 0.2	< 0.1	No
	GAC	5.4	5.4	4.9	5.0	Yes
	GAC	< 0.2	< 0.6	11.1	9.5	Yes
	NF/RO	0.6	< 0.6	0.6	< 0.6	No
	None	18.6	40.1	17.8	16.6	Yes
Surface Water	Alum	< 0.2	0.7	< 0.2	< 0.6	No
	Alum	1	1.6	1.8	2.3	Yes
	Alum	0.3	1.4	< 0.2	< 0.6	No
	Alum	7.6	5	< 0.2	6.1	Yes
	Ferric	<0.2	0.6	< 0.2	< 0.6	No

Tracing the levels and speciation patterns of chromium from source to tap was accomplished by collecting samples at treated water and distribution system locations in addition to the source water location. In most cases, the treated water was representative of the same source water supply post-treatment. However, in the case of the Fe/Mn facility targeted for Cr(VI) profiling, a second groundwater supply was blended with the source water supply prior to the treated water sampling location.

In general, the treated water chromium levels were comparable to that found for source waters with one important exception (Table 7-4). The alum plant that had Cr(III) as the dominant specie had nearly complete removal of the total chromium in its treated water sample with no increase in Cr(VI).

Table 7-4
Chromium speciation in treatment profile utilities

Source Type	Treatment Type[1]	Disinfection Type	Raw Water Levels, µg/L		Treated Water Levels, µg/L		Distribution System Levels, µg/L		Cr(VI) to Total Cr Ratio, %[2]		
			Cr(VI)	Tot Cr	Cr(VI)	Tot Cr	Cr(VI)	Tot Cr	Raw	Treated	Dist System
Ground	**Disinfection Only ****	**CL2**	**22.2**	**31.2**	**Not Sampled**		**0.4**	**< 0.1**	**71%**	**NS**	**> 100%**
	Disinfection Only	CL2	0.7	1.4	0.8	0.6	0.7	0.6	51%	135%	106%
	Disinfection Only **	**CL2**	**14.0**	**13.2**	**13.8**	**12.9**	**8.4**	**6.6**	**106%**	**107%**	**127%**
	Disinfection Only	CL2	1.3	2.2	4.2	4.2	11.9	11.4	62%	99%	104%
	Fe/Mn	CL2	< 0.2	0.4	< 0.2	0.1	< 0.2	2.8	0%	0%	0%
	Fe/Mn	CL2/CLM	< 0.2	< 0.1	< 0.2	< 0.1	0.4	0.4	I	I	95%
	Fe/Mn **	**CL2/CLM**	**< 0.2**	**0.1**	**< 0.2**	**< 0.1**	**< 0.2**	**0.3**	**0%**	**I**	**0%**
	GAC **	**CL2/CLM**	**11.3**	**9.3**	**10.8**	**10.2**	**< 0.2**	**0.2**	**122%**	**106%**	**0%**
	GAC **	**CL2**	**4.9**	**5.0**	**5.5**	**4.8**	**3.4**	**3.3**	**98%**	**114%**	**103%**
	NF/RO	CL2/CLM	0.6	0.3	< 0.2	< 0.1	Not Sampled		200%	I	NS
	None **	**NONE**	**17.8**	**16.6**	**Not Sampled**		**17.6**	**16.2**	**107%**	**NS**	**109%**
Surface	**Alum ****	**CL2**	**< 0.2**	**6.1**	**< 0.2**	**< 0.1**	**< 0.2**	**< 0.1**	**0%**	**I**	**I**
	Alum **	**CL2**	**1.8**	**2.3**	**1.5**	**1.3**	**1.7**	**1.6**	**80%**	**117%**	**109%**
	Alum	CL2	< 0.2	0.4	< 0.2	0.2	< 0.2	0.3	0%	0%	0%
	Alum	CL2/CLM	< 0.2	0.4	< 0.2	< 0.1	0.2	0.4	0%	I	55%
	Ferric	O3/CL2/CLM	< 0.2	0.3	< 0.2	0.1	< 0.2	0.1	0%	0%	0%

1. Those systems included in the treatment profile study to specifically capture the behavior of chromium are indicated by '**' and these results are shown in **BOLD**.
2. As appropriate to the treatment system, chromium speciation fractions were not determined when either (a) a location was not sampled because the intermediate location was not available to the system type (NS), or (b) the analytical results for both Cr(VI) and total chromium were below detection and the calculation of a speciation fraction was indeterminate (I).

Of the ten profiled systems with detectable source or treated water total and hexavalent chromium concentrations, six showed decreased concentrations in the distribution system, while three showed no change and one showed increased concentrations. Where changes in total and hexavalent chromium did occur between the treated and distribution system sampling locations, it appeared that the sampling location chosen in the distribution system was subject to blending from other source waters and not representative of the source water sampled for the treatment profile survey.

The results found from the treatment profile survey for hexavalent and total chromium indicated the following:

- Temporal variability in chromium concentrations should be expected for both surface and ground water supplies. This variation may include chromium speciation behavior.
- Hexavalent chromium was the dominant form of chromium in the potable water sources and treated supplies sampled during the treatment profile survey. However, the effects of oxidants on chromium speciation could not be distinguished from the treatment profile survey results. No reductants were used by the systems profiled.
- In the case of conventional treatment systems (coagulation with filtration), trivalent chromium appears to be effectively removed while hexavalent chromium is not removed.
- Other than the ability of conventional treatment to remove trivalent chromium, no other technology included in the profile survey affected hexavalent chromium or total chromium levels.

BORON TREATMENT PROFILE RESULTS

Ten water systems were specifically targeted for boron profiling through water treatment and distribution systems. Unlike with chromium, all of the systems participating in the treatment profile study had detectable levels of boron present in their source water (Table 7-5), enabling the tracking of boron from source-to-tap in all systems.

The surface and ground water supplies examined in the treatment profile study exhibited similar degrees of variation on average with relative differences between the two sampling events of 19% and 17% for ground and surface water supplies, respectively. This degree of variability is within analytical error bounds and does not generally represent significant variance.

Table 7-5
Comparison of source water sampling results for boron

Source Type	Treatment Type	National Occurrence Survey Source Boron, µg/L	Treatment Profile Source Boron, µg/L	Treatment Profile Targeted B
Ground	Disinfection Only	855	809.4	Yes
	Disinfection Only	2500	2276.9	Yes
	Disinfection Only	3320	2652.4	Yes
	Disinfection Only	140	132.8	No
	Fe/Mn	673	526.4	Yes
	Fe/Mn	354	201.8	Yes
	Ferric	163	120.2	Yes
	GAC	30.8	31.4	No
	GAC	128	192.7	No
	NF/RO	Not Sampled	1969	Yes
	None	68.5	66.8	No
Surface	Alum	345	405.1	Yes
	Alum	147	180.2	No
	Alum	343	323.2	Yes
	Alum	37.5	40.9	No
	Ferric	238	181.2	Yes

Boron was not effectively removed through the treatment systems investigated in the treatment profile survey (Figure 7-1). Even removal by the NF/RO groundwater system was negligible (15%). The maximum apparent removals ranged from 25% to 30% and occurred in a conventional surface water (alum) system and a groundwater system where iron/manganese treatment was practiced. These apparent removals are likely due to blending within the plants. Changes in boron concentration in the distribution system were also found to be negligible, and can also be explained by blending effects.

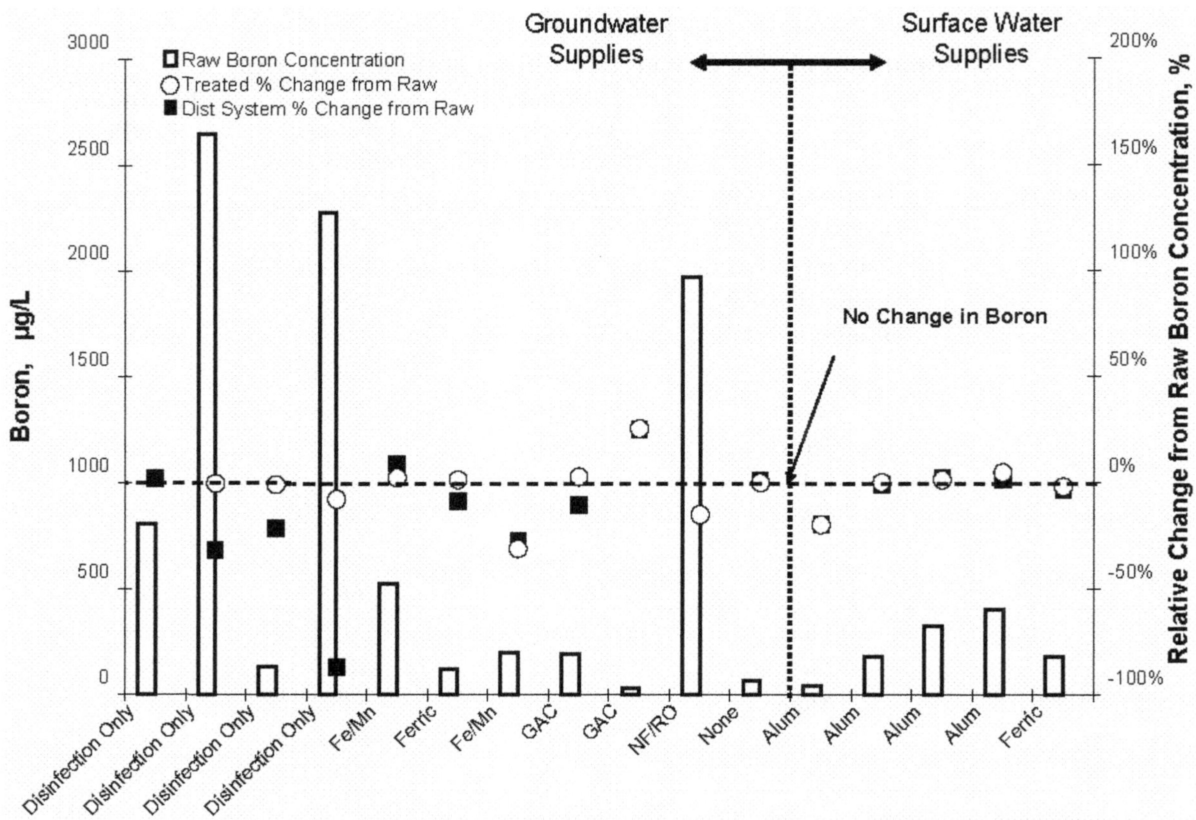

Figure 7-1. Treatment profile results for boron

Based on these results, boron levels in source waters would be expected to remain fairly conservative through most water systems. Other technologies that could affect boron concentrations at greater levels than those observed in this study may include lime softening and high pH membrane technologies. With few exceptions, source water boron represents a reasonable estimate of boron levels delivered to customers' taps.

CHAPTER 8
CONCLUSIONS

ANALYTICAL CHALLENGES

This project investigated analytical challenges for both chromium and boron analysis. The investigation focused on interferences, preservation techniques, and digestion methods for accurate and repeatable recovery performance. Laboratory experiments were first used to investigate the analytical challenges. Preliminary field sampling based on knowledge gained from the laboratory experiments validated the analytical methods used for the regular survey.

Interferences in boron analysis were found with aluminum and iron for at least one spectrum evaluated when background concentrations approached 100 mg/L as either Fe or Al. These concentrations are well above those expected in drinking water sources. All samples collected during this survey contained less than 100 mg/L Fe and Al; therefore, these interferences were not problematic.

Therefore at least two spectra were used to confirm all boron analyses. Boron was found to be otherwise unaffected by sample handling and processing methods as long as base preservation methods were employed (pH > 8).

Sampling for chromium necessitated consideration of both total chromium and hexavalent chromium analyses. Previous field sampling experiences had resulted in total chromium concentrations less than corresponding hexavalent chromium concentrations (Eaton et al. 2001). A series of lab experiments were conducted to determine the best approach for chromium analysis to correctly quantify both hexavalent and total chromium.

Using in-bottle acid digestion procedures that follow Method 3500-Fe B (APHA, AWWA, WEF 1998) – the hydroxylamine digestion to completely dissolve iron – was found to perform best in achieving complete recovery of total chromium. Base preservation for hexavalent chromium was found to be adequate when the sample pH was raised above pH 8.

The final survey sample kit preservation, digestion, and sample analysis procedures were developed based on results from the investigation of both chromium and boron analytical challenges. The final survey sample kit included two bottles: one (Bottle A) acid-preserved and the other (Bottle B) base-preserved. Sample kits were sent from Virginia Tech to participating drinking water utilities that collected samples following detailed sampling instructions. After samples kits were returned to Virginia Tech by the participating utilities, the acid-preserved sample was digested in the bottle with hydroxylamine following a procedure similar to that specified in Standard Method 3500 – Fe B (APHA, AWWA, WEF 1998). Following this procedure, 4% hydroxylamine and 2% hydrochloric acid were added to the bottle. The bottle was then heated to 85°C for 24 hours. The resulting digested solution was then filtered using a 0.45 µm pore size filter and analyzed for total boron by ICP-ES following Standard Method 3120 B (APHA, AWWA, WEF 1998) at Virginia Tech. An aliquot of the hydroxylamine digested and filtered sample was sent to Utah State University (USU) for total chromium analysis by ICP-MS following EPA Method 200.8 (Determination of Trace Elements in Waters and Wastes by Inductively Coupled Plasma – Mass Spectrometry) (USEPA 1994). The base-preserved sample was filtered in the Virginia Tech lab and then sent to USU for hexavalent chromium analysis by IC following EPA Method 1636 (Determination of Hexavalent Chromium by Ion Chromatography) (USEPA 1996a).

NATIONAL OCCURRENCE TRENDS

Several important results were identified from the NCBOS data. Boron was found to occur higher in groundwater sources than surface water sources and occurred equally across water system size categories. Source waters treated by NF/RO membranes exhibited the highest boron occurrence by treatment type, followed by MF/UF, softening, and disinfection only. No clear occurrence patterns exist by EPA region for groundwater boron occurrence; however, surface water boron occurs to a greater degree in the western U.S than in the eastern U.S.

A majority of the hexavalent chromium sample results were found to be less than the MDL of 0.2 µg/L. Total chromium was found to occur equally in surface waters and groundwaters; however, hexavalent chromium is not found in surface waters to nearly the same degree as in groundwaters. NCBOS total chromium concentrations in surface waters are – with few exceptions – composed primarily of trivalent chromium. Several groundwater results show total chromium concentrations composed exclusively by hexavalent chromium. This speciation trend has not been previously reported.

Both total and hexavalent chromium exhibited elevated concentrations in the medium 10,001 – 50,000 size category. This may be an artifact attributed primarily to source water type since the medium size category is dominated by groundwater sources.

The occurrence trends for total and hexavalent chromium are essentially the same for all treatment types except conventional treatment where total chromium concentrations are higher. Since these treatment types are typically associated with surface waters, this result is consistent with the chromium speciation occurrence trends by source water previously reported. This conclusion can also be made for waters treated by softening. The highest chromium concentrations are found in source waters not treated or just disinfected – typically associated with groundwaters.

When evaluating regional occurrence trends, chromium – both total and hexavalent – is greatest in EPA Region 9 which includes California. Total chromium occurrence is more variable by region in surface waters than groundwaters. No clear regional patterns exist for total chromium in surface waters. However, regional patterns did exist for groundwater results where total chromium occurred to the greatest degree in EPA Region 9.

TREATMENT PROFILE SURVEY FOR CHROMIUM AND BORON

The fate of chromium and boron in water systems was investigated by performing a special survey where longitudinal sampling was performed from source-to-tap in fifteen water systems.

The results found from the treatment profile survey for hexavalent and total chromium indicated the following:

- Temporal variability in chromium concentrations should be expected for both surface and ground water supplies. This variation may include chromium speciation behavior.
- Hexavalent chromium was the dominant form of chromium in the potable water sources and treated supplies sampled during the treatment profile survey. However, the effects of oxidants on chromium speciation could not be distinguished from the treatment profile survey results. No reductants were used by the systems profiled.

- In the case of conventional treatment systems (coagulation with filtration), trivalent chromium appears to be effectively removed while hexavalent chromium is not removed.
- Other than the ability of conventional treatment to remove trivalent chromium, no other technology included in the profile survey affected hexavalent chromium or total chromium levels.

Boron was not effectively removed through the treatment systems investigated in the treatment profile survey. Even removal by the NF/RO groundwater system was negligible (15%). The maximum apparent removals ranged from 25% to 30% and occurred in a conventional surface water (alum) system and a groundwater system where iron/manganese treatment was practiced. These apparent removals are likely due to blending within the plants. Changes in boron concentration in the distribution system were also found to be negligible, and can also be explained by blending effects.

Based on these results, boron levels in source waters would be expected to remain fairly conservative through most water systems. Other technologies with potential to remove boron concentrations to a greater degree than those observed in this study may include lime softening and high pH membrane technologies. With few exceptions, source water boron represents a reasonable estimate of boron levels delivered to customers' taps.

APPENDIX A:
RESULTS FROM EVALUATION OF SAMPLE PRESERVATION IN SIMULATED NATURAL WATERS

Boron

spikes:

B	100	ppb
Cr(III)	40	ppb
Cr(VI)	40	ppb

Legend:	
	1 = salt only
	2 = particulate
	3 = reducing
	4 = hard groundwater
	5 = chlorine
	6 = biologically active

Sample	Preserved	Filtered	Digestion	Boron Concentration (ppb) before spk	after spk	1 week	2 weeks
1	yes	yes	none	BDL	99.8	99.6	101.6
	yes	no	alk	BDL	97.0	98.2	96.3
	no	yes	alk	BDL	99.5	95.3	97.4
	no	no	alk	BDL	97.8	95.0	98.4
	no	no	acid	BDL	100.5	102.7	103.2
2	yes	yes	none	2.5	101.4	104.2	103.0
	yes	no	alk	4.0	93.0	103.4	102.7
	no	yes	alk	5.8	96.5	101.8	100.7
	no	no	alk	7.2	94.8	101.9	100.2
	no	no	acid	7.9	102.5	105.9	101.3
3	yes	yes	none	BDL	100.0	101.0	98.4
	yes	no	alk	3.4	97.9	95.0	87.3
	no	yes	alk	2.2	101.9	102.3	98.9
	no	no	alk	2.2	103.1	100.6	102.1
	no	no	acid	1.3	106.5	106.0	108.8
4	yes	yes	none	BDL	97.3	100.2	98.6
	yes	no	alk	2.3	92.7	97.9	96.3
	no	yes	alk	4.1	93.6	101.6	95.8
	no	no	alk	2.5	99.6	100.0	91.6
	no	no	acid	BDL	99.1	101.0	97.6
5	yes	yes	none	BDL	97.5	102.0	99.1
	yes	no	alk	1.9	90.6	100.1	94.3
	no	yes	alk	2.2	99.5	99.1	95.6
	no	no	alk	1.6	96.7	102.4	94.5
	no	no	acid	BDL	103.1	101.9	100.7
6	yes	yes	none	2.1	102.5	103.6	102.4
	yes	no	alk	4.1	102.5	101.7	93.6
	no	yes	alk	4.3	102.7	98.0	99.0
	no	no	alk	4.4	97.4	101.0	97.5
	no	no	acid	2.4	104.1	102.8	101.9

Total Chromium

spikes:

B	100 ppb	
Cr(III)	40 ppb	
Cr(VI)	40 ppb	

Legend:
1 = salt only
2 = particulate
3 = reducing
4 = hard groundwater
5 = chlorine
6 = biologically active

				Chromium Concentration (ppb)				
Sample	Preserved	Filtered	Digestion	Total Cr before spk	Total Cr after spk	Total Cr 1 week	Total Cr 2 weeks	USU Total (2 weeks)
1	yes	yes	none	BDL	57.3	46.4	42.2	47.7
	yes	no	alk	BDL	33.3	37.7	33.7	39.4
	no	yes	alk	BDL	42.1	34.4	35.4	39.2
	no	no	alk	BDL	41.0	34.8	35.8	33.4
	no	no	acid	BDL	81.0	73.5	69.8	62.4
2	yes	yes	none	1.2	37.7	41.6	41.2	40.7
	yes	no	alk	BDL	36.3	36.9	33.4	38.8
	no	yes	alk	BDL	BDL	BDL	BDL	0.2
	no	no	alk	BDL	35.2	16.2	BDL	7.5
	no	no	acid	1.0	74.5	81.1	23.8	25.8
3	yes	yes	none	BDL	48.3	48.6	53.2	51.5
	yes	no	alk	BDL	53.7	47.8	51.5	53.7
	no	yes	alk	BDL	42.0	29.3	34.3	34.8
	no	no	alk	BDL	52.3	37.7	41.8	31.7
	no	no	acid	BDL	80.3	71.5	78.0	67.5
4	yes	yes	none	BDL	47.0	36.7	38.8	38.5
	yes	no	alk	BDL	32.2	33.7	33.0	35.9
	no	yes	alk	BDL	30.7	34.6	34.7	37.0
	no	no	alk	BDL	33.4	34.0	35.3	24.9
	no	no	acid	BDL	77.3	82.0	78.5	70.3
5	yes	yes	none	BDL	48.0	42.6	65.7	66.5
	yes	no	alk	BDL	32.2	52.3	50.3	59.2
	no	yes	alk	BDL	42.1	66.5	53.8	65.6
	no	no	alk	BDL	49.0	72.0	72.8	60.3
	no	no	acid	BDL	82.0	81.6	82.3	81.4
6	yes	yes	none	BDL	59.4	61.5	59.2	59.1
	yes	no	alk	BDL	33.5	45.4	36.5	46.4
	no	yes	alk	BDL	32.6	34.1	32.7	35.2
	no	no	alk	BDL	31.9	37.6	32.2	35.7
	no	no	acid	BDL	76.3	64.3	61.8	66.0

Hexavalent Chromium

spikes:	B	100 ppb
	Cr(III)	40 ppb
	Cr(VI)	40 ppb

Legend:	
	1 = salt only
	2 = particulate
	3 = reducing
	4 = hard groundwater
	5 = chlorine
	6 = biologically active

				Chromium Concentration (ppb)				
Sample	Preserved	Filtered	Digestion	Cr(VI) before spk	Cr(VI) after spk	Cr(VI) 1 week	Cr(VI) 2 weeks	USU Cr(VI) (2 weeks)
1	yes	yes	none	BDL	41.7	36.2	41.0	41.4
	yes	no	alk	BDL	30.4	39.2	33.0	33.1
	no	yes	alk	BDL	39.2	39.4	33.8	34.9
	no	no	alk	BDL	38.8	38.5	33.6	33.9
	no	no	acid					
2	yes	yes	none	BDL	44.5	49.9	44.3	41.0
	yes	no	alk	BDL	34.4	44.2	37.9	37.5
	no	yes	alk	BDL	BDL	BDL	BDL	0.7
	no	no	alk	BDL	35.6	22.0	15.7	4.1
	no	no	acid					
3	yes	yes	none	BDL	42.6	44.9	47.7	49.6
	yes	no	alk	BDL	54.3	51.0	44.5	46.6
	no	yes	alk	BDL	40.6	34.3	31.7	29.7
	no	no	alk	BDL	51.4	44.3	37.6	37.1
	no	no	acid					
4	yes	yes	none	BDL	41.6	33.2	40.2	39.7
	yes	no	alk	BDL	30.6	37.8	32.3	32.2
	no	yes	alk	BDL	29.3	38.1	34.9	34.3
	no	no	alk	BDL	32.8	37.9	34.3	34.8
	no	no	acid					
5	yes	yes	none	BDL	43.7	41.1	47.4	45.1
	yes	no	alk	BDL	32.4	40.3	37.5	38.2
	no	yes	alk	BDL	39.9	73.8	61.7	62.7
	no	no	alk	BDL	41.6	68.6	62.7	65.7
	no	no	acid					
6	yes	yes	none	BDL	42.5	43.0	41.9	40.4
	yes	no	alk	BDL	29.9	37.9	32.6	33.2
	no	yes	alk	BDL	32.8	42.2	32.0	32.8
	no	no	alk	BDL	31.4	35.7	31.8	32.7
	no	no	acid					

APPENDIX B:
FIELD SAMPLING PROTOCOL FOR THE PRELIMINARY SURVEYS

This appendix includes a detailed listing of the bottles to be included in each preliminary survey sample kit (Attachment 1, Table 1). This includes information on preservatives to be added to each bottle before mailing and how each bottle will be handled after it is returned. The cover letter and a detailed instruction sheet to be included in each kit that is mailed is also included.

Table B.1. Listing of bottles A-H to be included in the initial 20 plant survey.

Preliminary 20 Plant QA/QC Survey
Description of Bottles to be Included in Sampling Kits

Stock Solutions
Boron Spike Solution: 10 mg B / L
Chromium(VI) Spike Solution: 10 mg Cr(VI) / L
Chromium(III) Spike Solution: 10 mg Cr(III) / L
Soda Ash Solution: 100 g Na_2CO_3 / L

Treatment before sending bottles------------------------------->					In the Field---->		Treatment in Lab---------------------------->				
Bottle	Preservative	Amount of Preservative (mL)	Spike	Spike[&] Amount (mL)	Sample Filtered in Field	Approx. Sample (mls)	Digestion in Lab	Filter Back in Lab	Analysis for	Analysis by	Goal
A	Nitric Acid (pH < 2)	0.333	none		no	200	Acid* Acid*	Yes, after digestion	B Cr	ICP-ES GFAA	Total B Total Cr
B	Nitric Acid (pH < 2)	0.167	B Cr(VI)	1 0.1	no	100	Acid[@] Acid[@]	Yes, after digestion	B Cr	ICP-ES GFAA	Insure recovery of spike Insure recovery of spike
C	12 mM Soda Ash	1.27	none		no	100	none	Yes	Cr(VI)	IC	Recoverable Cr(VI)
D	12 mM Soda Ash	1.27	Cr(VI)	0.1	no	100	none	Yes	Cr(VI)	IC	Insure recovery of spike
E	12 mM Soda Ash	0.38	none		yes, into bottle	30	none Acid[@] Acid[@]	No No No	Cr(VI) B Cr	IC ICP-ES GFAA	Soluble Cr(VI) Soluble B Soluble Cr (3+6)
F	12 mM Soda Ash	0.38	B Cr(III)	0.3 0.03	yes, filtered into bottle	30	none Acid[@] Acid[@]	No No No	Cr(VI) B Cr	IC ICP-ES GFAA	Insure recovery of spike Insure recovery of spike Insure recovery of spike
G	100 mL of laboratory reagent-grade water with 12 mM soda ash preservative										
H	empty										

Notes
* Sampling Protocol for this bottle is detailed on next page
@ Acid Digestion = Std. Method 3030E; 5% nitric acid, 105 C, 2 hours)
& Spikes will be 100 ppb boron and 10 ppb Cr(VI) or 10 ppb Cr(III)

Cover letter and sampling instructions.

To whom it may concern:

Thank you for agreeing to participate in the American Water Works Association Research Foundation (AwwaRF) survey of boron and chromium occurrence in surface and groundwaters throughout the United States. We have included a sampling kit and detailed instructions. Please read over the sampling instructions before collecting the sample, and make sure that your kit is complete. If there is any problem with the kit or if you have a problem collecting a sample, please give us a call at (540) 231-2516, leave your name and phone number for Mr. Jeff Parks and he will call you back to determine how to proceed. Or send an e-mail to "jparks@vt.edu". He will respond promptly with answers to any questions.

Collect the samples as described in the instructions within 2 weeks after receiving this kit, place the sample bottles into the self-addressed return container (we have already included the appropriate postage), and put it in the US mail so that we (Virginia Tech) can receive the samples.

Thank you for participating in this survey. The results will provide valuable insights to possible future regulation of boron and chromium in US drinking water.

Regards,

Jeffrey Parks

Virginia Tech

INSTRUCTIONS FOR USE OF SAMPLING KIT

Important: Many bottles have a liquid preservative inside them. Try to prevent loss of the preservative during sampling. Also, some preservatives are 2% concentrated nitric acid, and some are liquid soda ash (Na_2CO_3). If any of these preservatives should inadvertently contact you, rinse the contacted area thoroughly with clean water.

Your boron/chromium sampling kit includes the following items:

1) Three 30 ml plastic syringes in package
2) Three screw-on filter attachments
3) 8 Pre-labeled sample bottles (1-250 mls and 7-120 mls each). Bottles are labeled "A" - "H."
4) Gloves
5) Return envelope (pre-addressed and with correct postage affixed)

The overall strategy is as follows. You will first separate all bottles marked "A" through "F" from the package and then collect filtered and unfiltered samples of raw influent water. Thereafter, you will filter a portion of 100 ml of laboratory reagent-grade water (supplied in bottle "G") into bottle "H." Analysis of that sample will serve as a control sample to evaluate possible contamination. Finally, you will ship the bottles back to us for analysis.

<u>Detailed Steps</u>

0) Check to make sure you have all items listed above.

1) You will also need a ***clean*** plastic sampling container (beaker or flask) of about 1 L (1000 mls) size. Glass cannot be used since it can contaminate samples with boron.

2) Go to the sample location where you can collect influent water samples.

3) Put on gloves.

4) Taking care not to let dirt or other contamination into your plastic sampling container, rinse the container three times in the water to be sampled. Collect the sample by filling the container. Remove the cap from bottle "A" and place the cap upside-down in a clean location (this ensures that no dirt gets inside the cap to cause contamination of the sample).

5) After swirling the water in your container to mix contents thoroughly, pour your sample directly from the container into bottle "A" and fill the bottle approximately to the blue line (1" from the top). Replace cap on the bottle (tightly)! It does not matter if the sample is slightly above or below the blue line.

6) Repeat step 5 for bottles "B", "C", and "D", filling them up to the blue line.

7) Collect a filtered sample of the raw water in bottle "E" by:

a) Rinse and fill the syringe with your raw water. In this step, it is OK if solids are on the bottom of your sample container, so you do not have to swirl the contents of the sample container before sampling. Take the syringe out of its package and remove the blue plastic tip from the syringe. Using a sucking motion (by pulling back the handle on the syringe) while immersing the tip of the syringe under the surface of the water sample, fill the enclosed 30 ml syringe with your raw water sample and discharge onto the ground to rinse syringe. Completely fill the syringe once more with the raw water by drawing back on the handle.

b) Screw on filter attachment and filter the sample. Take a filter from the package and screw clockwise until snug onto the end of the syringe (which is still full with your sample). While pointing the filter upwards, gently push about 8 drops of sample through the filter to remove entrained air and partly rinse the filter. Then firmly push the sample through the filter and into bottle "E." This may take 30 seconds or up to two minutes depending on how dirty your sample is. When many solids are present, it is possible that the filter will become clogged and flow will decrease to less than 2 drops/sec. In this case, just push as much water through the filter as possible in 3-5 minutes or stop after 30 ml have been filtered. Replace cap on the bottle (tightly)!

8) Using a new syringe and filter, repeat step 7 for bottle "F" instead of bottle "E."

9) Using a new syringe and filter, and the water in sample bottle "G" as the source of water, filter 30 ml of the sample from bottle "G" into bottle "H."

10) Make sure that all caps are tightly closed on all bottles.

11) Please fill out the short survey that is attached to these instructions. Place the completed survey and all the sample bottles into the return envelope and seal carefully. We have attached sufficient postage for first class delivery by US mail. Dispose of other sampling kit materials in the normal garbage.

Survey

1) Was there a rotten egg odor (sulfides) in the raw water? Yes or No

2) Does your raw water sometimes contain soluble iron or manganese at levels of concern to consumers (> 0.05 mg/L soluble iron or Mn^{+2})? Yes or No

3) What raw water source/mixture were you using on the date sampled?____

4) What is the date samples were collected?__________________________________

5) Name of person collecting sample____________________________________

6) Approximate pH of raw water (if known)? __________________________________

7) Approximate turbidity of raw water (if known)? __________________________________

8) Approximate alkalinity of raw water (if known)? __________________________________

9) Are there any metallic plumbing materials (i.e., copper, iron, stainless steel) that contact your raw water sample before it is collected? Yes or No.

10) Is oxidant or disinfectant (ozone, chloride dioxide, chlorine, chloramines, or permanganate) added to your raw water? Yes or No.

Thank you for completing the survey!

APPENDIX C:
SAMPLE KIT LETTER, INSTRUCTIONS, AND QUESTIONNAIRE FOR THE REGULAR SURVEY

Virginia Tech
VIRGINIA POLYTECHNIC INSTITUTE AND STATE UNIVERSITY

The Charles Edward Via, Jr. Department of Civil And Environmental Engineering
Environmental Engineering Program
418 Durham Hall
Blacksburg, Virginia 24061-0246
(540) 231-6131 or (540) 231-4595
Fax: (540) 231-7916

Sampling Coordinator
ABC Water Utility
123 Main Street
Anywhere, CO 80202

Utility ID: 1

Dear Sampling Coordinator,

Thank you for participating in AwwaRF Project 2759 – an occurrence survey of chromium and boron in drinking water sources throughout the United States. This project team includes McGuire Environmental Consultants, Inc., Virginia Tech, and Utah State University.

This package includes sampling kits for each water source that you identified when signing up to participate in the survey project. The sources you are to sample include:

Source ID	Source Name	Source Type
1	Surface Water Source	Surface
2	Groundwater Source	Ground

Detailed sampling instructions and a written sampling questionnaire follow this cover letter. Samples should be collected *within 2 weeks* after receiving this kit. On the day you choose to sample your specified source waters, read the attached sampling instructions carefully and make sure that you understand them. Collect the samples as described in the instructions, complete the sampling questionnaire, place the samples and questionnaire in the provided pre-addressed return shipping carton, and mail the package to Virginia Tech.

If you have any questions regarding this survey or the sample instructions, please contact Chad Seidel of McGuire Environmental Consultants, Inc. (Phone: 303.623.0122; Email: cseidel@mcguireinc.com). He will respond promptly to answer your question. Jeff Parks may also be contacted by email (jparks@vt.edu) with questions regarding sampling.

Thank you for participating in this survey. We are sure that these water samples will help us better understand boron and chromium occurrence throughout the country.

Regards,

Jeffrey Parks
Ph.D. Candidate - Virginia Tech

SAMPLING KIT INSTRUCTIONS

Your source water sampling kit(s) includes the following items for each source to be sampled:

1) Pre-labeled sample bottles (125 mL). Bottles are labeled with your Utility ID, the source ID, and "A" and "B".

Important: Each bottle contains a small amount of liquid preservative. Bottle A contains a strong acid preservative and Bottle B contains a strong base preservative. Make sure that none of the liquid preservative already present in the bottle is lost while collecting the samples. Treat the bottles with appropriate precautions.

2) Gloves
3) Return packaging
4) Sampling questionnaire for each source water

The general sampling strategy for each source is as follows (detailed instructions are provided below). You will collect unfiltered samples of raw source water and add to each sample bottle. After sample collection, you will ship the bottles back to Virginia Tech for analysis.

<u>Detailed Steps</u>

0) Check to make sure you have all items listed above.

1) You will also need a ***clean*** plastic container (beaker or flask) of about 1 L (1000 mls) size to help in collecting the samples. Glass containers cannot be used since it could contaminate samples with boron.

2) Go to the specified source water location where you can collect a raw water sample.

3) Put on gloves.

4) Taking care not to let dirt or other contamination into your 1L plastic container, rinse the container twice in the water to be sampled. Then, collect a water sample by filling the container. Remove the cap from bottle "A" and place it upside-down in a clean location (to ensure that no dirt gets inside the cap and cause contamination of the sample later).

5) After swirling the water in your 1L container to mix contents thoroughly, pour your sample directly from the container into bottle "A" and fill the bottle approximately to the blue line (1" from the top). IT IS BEST IF YOU DO NOT FILL BOTTLE COMPLETELY. Replace the cap on the bottle (tightly)! It does not matter if the sample is slightly above or below the blue line.

6) Repeat step 5 for bottle "B".

7) Make sure that both caps are tightly closed on both bottles.

8) Please fill out the short questionnaire that is attached to these instructions. Place the completed questionnaire and the sample bottles into the shipping package provided and seal carefully. Affix the appropriate amount of postage and mail back to Virginia Tech (each package should weigh less than 1 lb. and cost $3.85 from anywhere in US if mailed by USPS Priority Mail).

SOURCE SAMPLING QUESTIONNAIRE

Utility Name: ABC Water Utility
Utility ID: 1

Please answer the following questions for each individual source identified.

Source Name	**Surface Water Source**	**Groundwater Source**			
Source Type	**Surface**	**Ground**			
Source ID	**1**	**2**			
Sampling date and time?					
Person collecting sample?					
pH of sampled water (if known)?					
Approximate turbidity of sampled water (if known)?					
Does the sampled water often contain soluble iron or manganese at levels of concern to consumers (> 0.05 mg/L soluble iron or Mn^{+2})? Yes or No					
Are there any metallic plumbing materials that come in contact with the source water sample before it was collected? Yes or No					
How is this drinking water source treated?					
Was any treatment performed upstream of the sampling location (e.g. pre-oxidation, pre-disinfection)?					

Thank you for participating in the AwwaRF Chromium and Boron Occurrence Survey!

APPENDIX D:
TREATMENT PROFILE SAMPLING LOCATION DIAGRAMS

City of Davis, California

- Utility ID: 11
- Source Name: All Wells

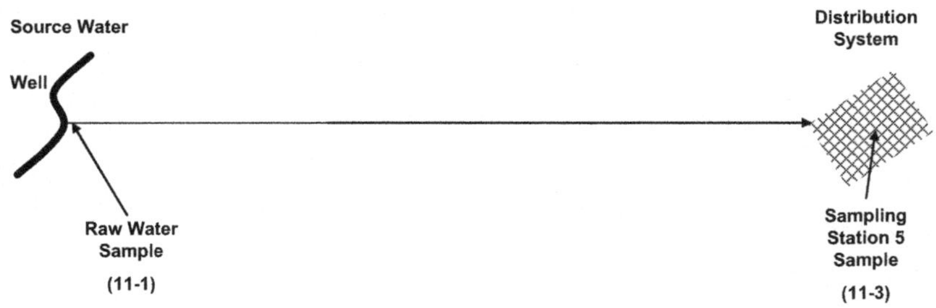

Goleta Water District

- Utility ID: 32
- Source Name: Lake Cachuma

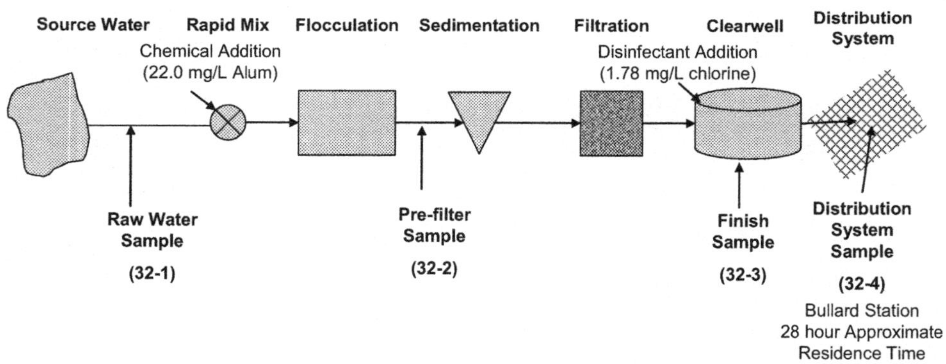

North Edwards Water District

- Utility ID: 33
- Source Name: Well #2 (Secondary)

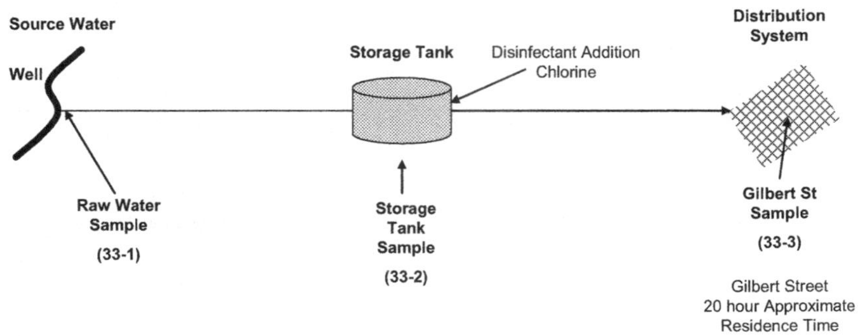

Oak Trail Ranch Mutual Water Co.

- Utility ID: 37
- Source Name: Well 1

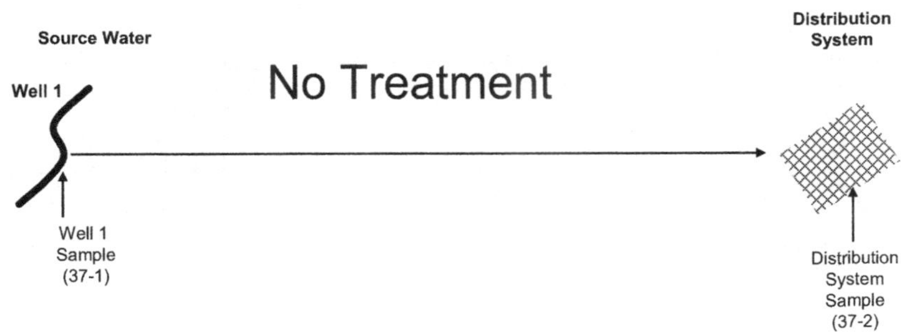

City of Lodi

- Utility ID: 48
- Source Name: Lodi Well No.18

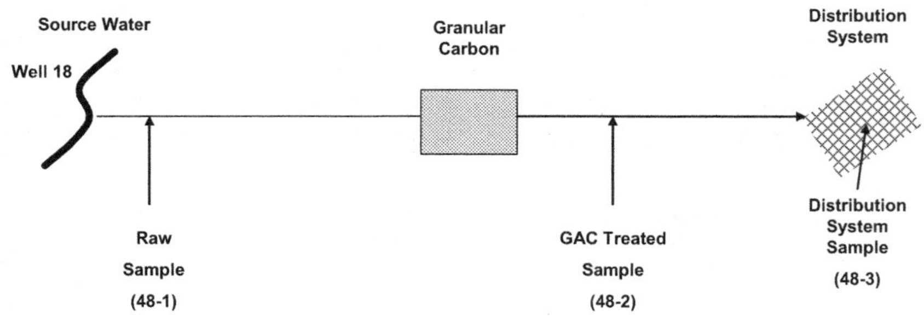

City of Tempe

- Utility ID: 52
- Source Name: South Treatment Plant

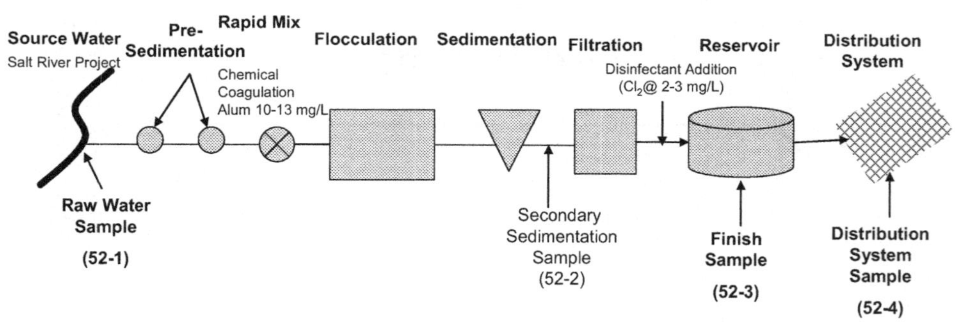

Glendale Water & Power

- Utility ID: 77
- Source Name: Glendale WTP

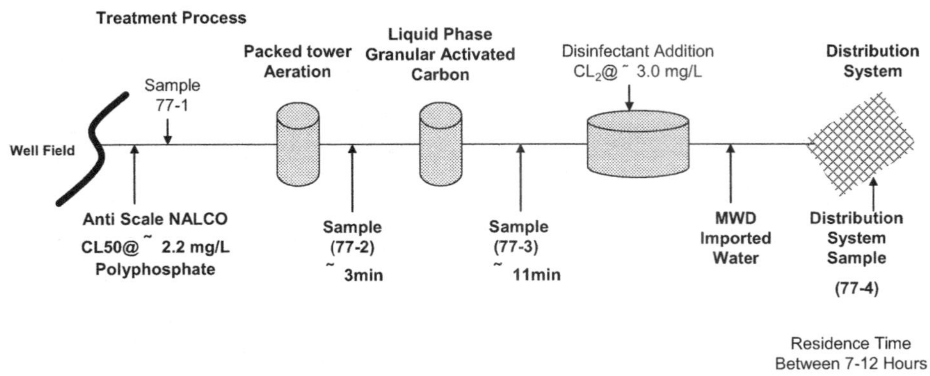

City of Midland

- Utility ID: 78
- Source Name: 1

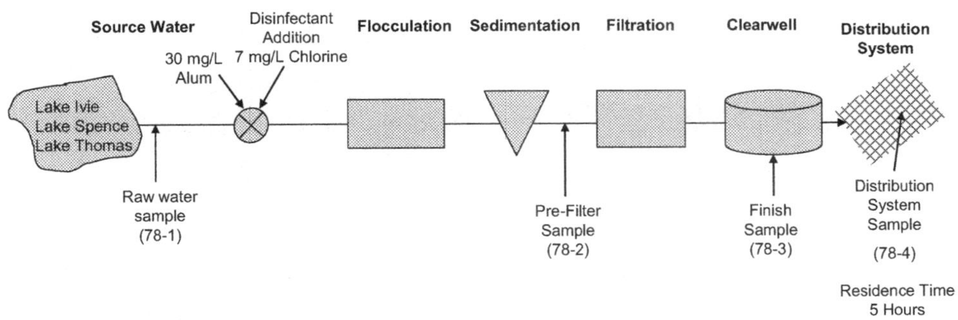

Kauai Dept. of Water – Wailua/Kapaa System

- Utility ID: 91
- Source Name: Waipao Valley

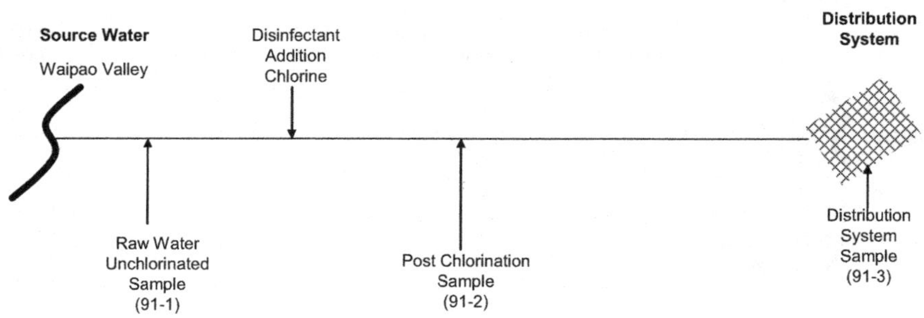

Castaic Lake Water Agency

- Utility ID: 106
- Source Name: Castaic Lake

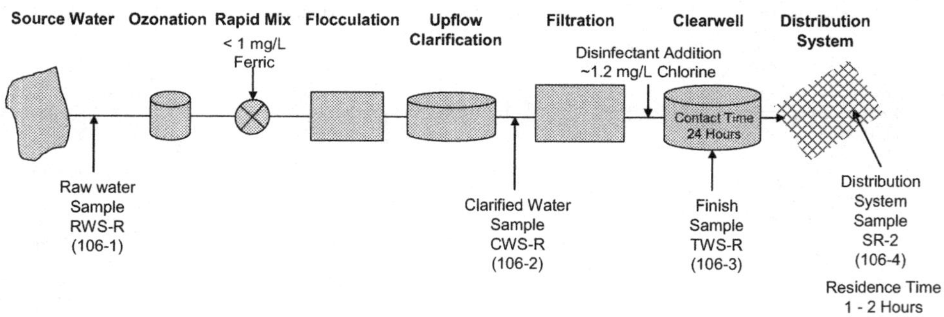

Castaic Lake Water Agency – Santa Clarita Division

- Utility ID: 108
- Source Name: Lost Canyon 2

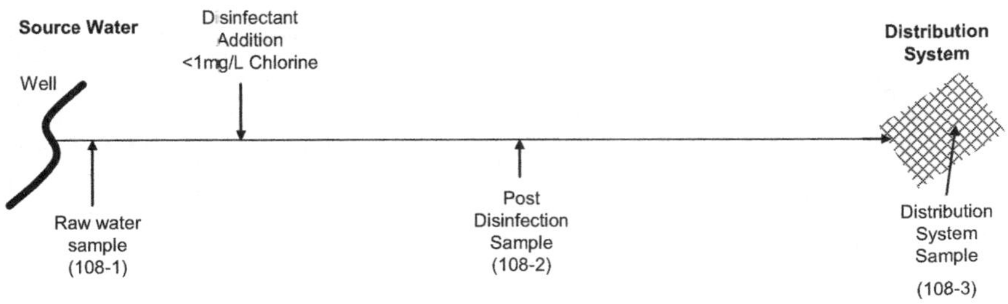

Jefferson Parish Water Dept.

- Utility ID: 114
- Source Name: Mississippi River

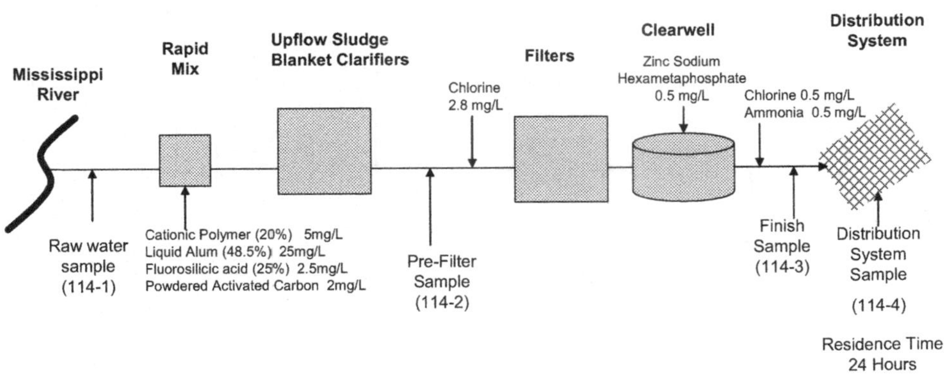

Newburgh Operation & Treatment Center

- Utility ID: 143
- Source Name: Newburgh

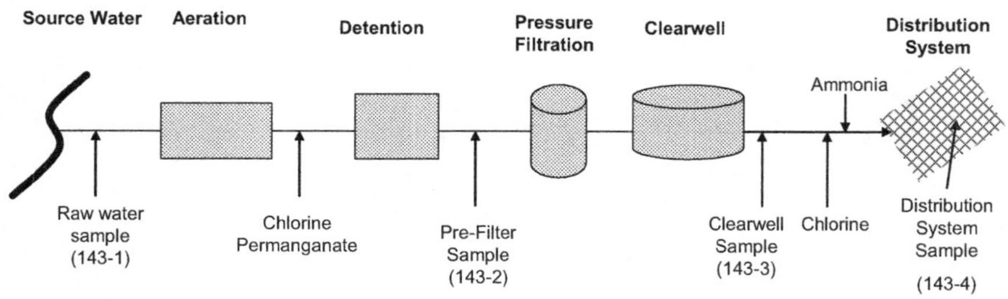

Indiana American – Terre Haute

- Utility ID: 160
- Source Name: Collector Well & Well 2

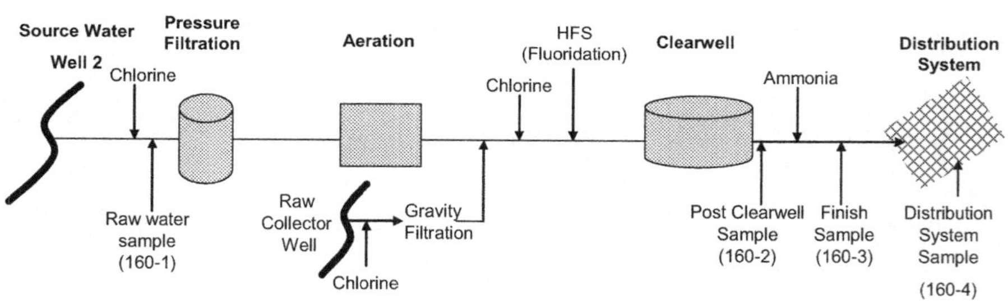

Indiana-American Water Co., Richmond District

- Utility ID: 164
- Source Name: South 4th Street Plant

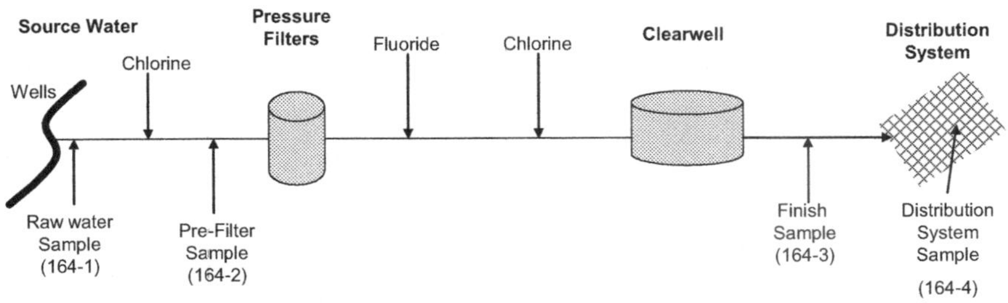

Newport News Water Works

- Utility ID: 184
- Source Name: BGD Feed

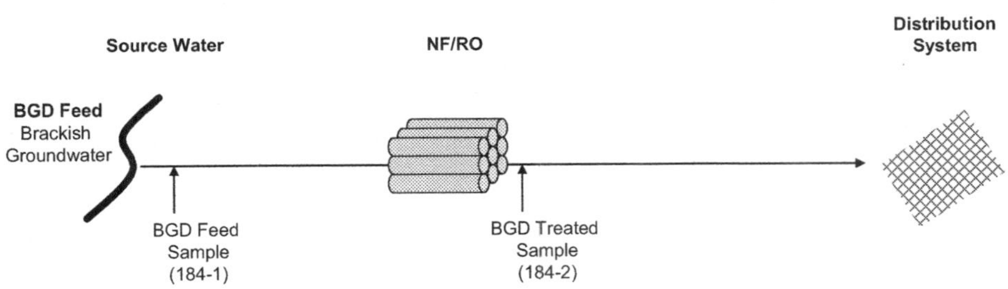

REFERENCES

APHA, AWWA and WEF (American Public Health Association, American Water Works Association, and Water Environment Federation). 1998. *Standard Methods for the Examination of Water and Wastewater*. 20th ed. Washington, D.C.: APHA.

Baes, C. F., Jr., and R. E. Mesmer. 1976. *The Hydrolysis of Cations.* John Wiley and Sons, New York, New York.

Bailey, R.P., and M.M. Benjamin, 1992. Sorption onto and Recovery of Hexavalent Chromium using Iron-Oxide Coated Sand. *Water Science and Technology*, (26) 5-6: 1239-1244.

Brandhuber, P., M. Frey, M. J. McGuire, P. Chao, C. Seidel, G. Amy, J Yoon, L. McNeill, K. Banerjee. 2004. *Treatment Options for Low-Level Hexavalent Chromium Removal Tested at Bench Scale*, Denver, CO: American Water Works Association Research Foundation.

Butler, L. N. 1967. *Ionic Equilibrium: A Mathematical Approach*. New York: Addison-Wesley.

California Department of Health Services. 2003. Drinking Water Program, Chromium-6 Monitoring and Regulation Webpage, http://www.dhs.ca.gov/ps/ddwem/chemicals/Chromium6/Cr+6index.htm

Coughlin, J. R. 1998. "Sources of Human Exposure: Overview of Water Supplies as Sources of Boron." *Biological Trace Element Research* 66: 87-100.

Clifford, D., and J. Man Chau. 1987. The Fate of Chromium (III) in Chlorinated Water. EPA/600/2-87/100. Washington, D.C.: US EPA.

Dean, J. A., Ed. 1987. *Lange's Handbook of Chemistry*. New York, McGraw-Hill Book Company.

Downing, R. G., P. L. Strong, B. M. Hoanec and J. Northington. 1998. Considerations in the Determination of Boron and Low Concentrations. *Biological and Trace Elements Research*. Vol. 66, No. 1-3, p. 3-21.

Durfor, C.N. and E. Becker. 1964. *Public Water Supplies of the 100 Largest Cities in the United States*. USGS Water Supply Paper, 1812:364.

Dzombak D. A. and F. M. M. Morel. 1990. *Surface Complexation Modeling: Hydrous Ferric Oxide*. John Wiley & Sons, New York.

Eaton, A., L. Ramirez, and A. Haghani. 2001. The Erin Brockovich Factor – Analysis of Total and Hexavalent Chromium in Drinking Waters. American Water Works Association Water Quality and Technology Conference Procedings, Nashville, TN.

Edwards, M, and M.M. Benjamin, 1989. Regeneration and Re-use of Iron Hydroxide Absorbents in Treatment of Metal-Bearing Wastes. *J. WPCF*, (61) 4: 481-490.

Goulden, P.D. and Y. P. Kakar. 1976. Determination of boron in high levels of nitrate and organics. *Water Research*, 10: 491-495.

Hodgman, C.D., R.C. Weast, and S.M. Selby, eds. 1961. *Handbook of Chemistry and Physics*, 43rd ed. Cleveland, Ohio: Chemical Rubber Publishing Company.

Holleman, A. F. and E. Wiberg. 2001. *Inorganic Chemistry*. New York, Academic Press.

Kaczynski, S.E., and R.J. Kieber. 1993. Aqueous Trivalent Chromium Photoproduction in Natural Water. *Environmental Science and Technology*, 27(8):1572–1576.

Kharkar, D. P., K. K. Turekian, and K. K. Bertine. 1968. Stream supply of dissolved silver molybdenum, antimony, selenium, chromium, cobalt, rubidium and cesium to the oceans, *Geochimica et Cosmochimica Acta*, 32: 285-298.

Lapp, T. W. and G. R. Cooper. 1976. Chemical Technology and Economics in Environmental Perspectives; Task II - Removal of Boron from Wastewater. Kansas City, MO, Midwest Research Institute (for EPA Office of Toxic Substances).

Longtin, J.P. 1988. Occurrence of radon, radium, and uranium in groundwater: *Journal of the American Water Works Association*. p. 84–93.

Mackin, J. E. 1986. The free-solution diffusion coefficient of boron: influence of dissolved organic matter. *Marine Chemistry*. 20(2): 131-140.

Nriagu, J.O. and E. Nieboer. 1988. *Chromium in the Natural and Human Environments*. New York, NY: John Wiley & Sons.

OEHHA. 1999. Public Health Goal for Chromium in Drinking Water. California Environmental Protection Agency.

Parks, J. L., L. McNeill, M. Frey, A.D. Eaton, A. Haghani, L. Ramirez, M. Edwards. 2004. Determination of total chromium in environmental water samples. *Water Research*, 39: 2827 – 2838.

Pettine, M., L. Campanella, F. Millero. 2002. Reduction of hexavalent chromium by H_2O_2 in acidic solutions. *Environmental Science and Technology*, (36) 5: 901-907.

Power, P. P. and W. G. Woods. 1997. The chemistry of boron and its speciation in plants. *Plant and Soil* 193: 1-13.

Sass, B. and D. Rai. 1987. Solubility of Amorphous Chromium(III)-Iron(III) Hydroxide Solid Solutions. *Inorganic Chemistry*, 26: 2228-2232.

Sengupta, A., D. Clifford, and S. Subramonian. 1986. Chromate Ion-Exchange Process at Alkaline pH. *Water Research*, 20(9):77–84.

Shirodkar, P. V. and S. Y. S. Singbal. 1992. Boron chemistry in relation to its variations in eastern Arabian Sea. *Indian Journal of Marine Sciences*. 21: 178-182.

Smith, B. M., P. Todd, et al. 1995. Boron Removal by Polymer-Assisted Ultrafiltration. *Separation Science and Technology*. 30(20): 3849-3859.

Tong, J.Y., and E.L. King. 1953. A Spectrophotometric Investigation of the Equilibria Existing in Acidic Solutions of Chromium(VI). *Jour. American Chemical Society*, 75:1902–1906.

USEPA. 1994. Methods for the determination of metals in environmental samples, supplement I: EPAP/600/R-94/111. US EPA, Washington, D.C.

USEPA. 1996a. Method 1636: Determination of Hexavalent Chromium by Ion Chromatography, EPA 821-R-96-003. US EPA, Washington, D.C.

USEPA. 1996b. Method 1639: Determination of Trace Elements in Ambient Waters by Stabilized Temperature Graphite Furnace Atomic Absorption, US EPA, Washington, D.C.

USEPA. 2000. Method 3060A: Alkaline Digestion for Hexavalent Chromium. *SW-846: Test Methods for Evaluating Solid Waste, Physical/Chemical Methods*, US EPA, Washington, D.C.

USEPA. 2004. FACTOIDS: Drinking Water and Ground Water Statistics for 2003. US EPA, Washington, D.C.

Waggott, A. 1969. An Investigation of the Potential Problem of Increasing Boron Concentrations in Rivers and Water Courses. *Water Research.* 3: 749-765.

Weast, R.C., M.J. Astle, and W.H. Beyer, eds. 1988. CRC Handbook of Chemistry and Physics, 69th ed. Boca Raton, Fla.: Chemical Rubber Company.

Wood, John and K. Nicholson. 1995. Boron determination in water by ion-selective electrode. *Environmental International*, (21) 2: 237-243.

Woods, W. G. 1996. "Review of Possible Boron Speciation Relating to Its Essentiality." The Journal of Trace Elements in Experimental Medicine 9:153-163.

ABBREVIATIONS

16-State	USEPA 16-State occurrence data
BDL	below detection limit
°C	degrees Celsius
$CaCO_3$	calcium carbonate
CalDHS	California Department of Health Services
Cl_2	chlorine
CLM	chloramine
Cr	chromium
Cr(III)	trivalent chromium
Cr(VI)	hexavalent chromium
CrO_4^{2-}	chromate
$Cr_2O_7^{2-}$	dichromate
$Cr(OH)_3$	chromium (III), hydroxide
DHS	(California) Department of Health Services
Fe	iron
Fe(II)	ferrous iron, iron (II)
Fe(III)	ferric iron, iron (III)
$FeO•Cr_2O_3$	chromite
GAC	granular activated carbon
GW	groundwater
HBN	Hydrologic Benchmark Network
HCl	hydrochloric acid
HCO_3	bicarbonate
$HCrO_4^-/CrO_4^{2-}$	chromate
HDPE	high-density polyethylene
HNO_3	nitric acid
hr	hour, hours
IC	ion chromatography
ICP–ES	inductively coupled plasma–emission spectrometer
ICP–MS	inductively coupled plasma–mass spectrometer
L	liter
M	molar
MCL	maximum contaminant level
MCLG	maximum contaminant level goal
MDL	method detection limit

MF/UF	microfiltration/ultrafiltration
Mg	magnesium
MGD	millions of gallons per day
mg/L	milligram per liter
min	minute
mL	milliliter
mM	millimolar
Mn	manganese
µg/L	microgram per liter
Na	sodium
NaCl	sodium chloride
Na_2CO_3	sodium carbonate, soda ash
$NaHCO_3$	sodium bicarbonate
NaOH	sodium hydroxide
NASQAN	National Stream Quality Accounting Network
NAWQA	National Water Quality Assessment
NCBOS	National Chromium and Boron Occurrence Survey
NF/RO	nanofiltration/reverse osmosis
NIRS	National Inorganic and Radionuclide Survey
NOM	natural organic matter
NWIS	National Water Information System
OEHHA	Office of Environmental Health Hazards Assessment
pE	negative logarithm of apparent electron activity; indicator of oxidation–reduction (redox) potential
PHG	public health goal
redox	reduction–oxidation
SDWIS	Safe Drinking Water Information System
STORET	STOrage and RETrieval
SW	surface water
TDS	total dissolved solids
USEPA	U.S. Environmental Protection Agency
USGS	U.S. Geological Survey
UV_{254}	ultraviolet absorbance at 254 nanometers
WQN	Water-Quality Monitoring Networks

IWA Publishing's authorised EU representative for General Product Safety Regulations is Diane D'Arras, 15 rue Duret, 75116 Paris, France, e-mail: safety@iwap.co.uk.

Printed and bound by CPI Group (UK) Ltd, Croydon, CR0 4YY
17/09/2026
02218798-0011